DATE DUE

DEMCO 38-297

The Chemistry of
Chalcones and
Related Compounds

The Chemistry of Chalcones and Related Compounds

DURGA NATH DHAR

Indian Institute of Technology
Kanpur, India

Foreword by

SIR DEREK BARTON

Gif-sur-Yvette, France

A Wiley-Interscience Publication

JOHN WILEY & SONS, New York • Chichester • Brisbane • Toronto

Copyright © 1981 by John Wiley & Sons, Inc.

All rights reserved. Published simultaneously in Canada.

Reproduction or translation of any part of this work
beyond that permitted by Sections 107 or 108 of the
1976 United States Copyright Act without the permission
of the copyright owner is unlawful. Requests for
permission or further information should be addressed to
the Permissions Department, John Wiley & Sons, Inc.

Library of Congress Cataloging in Publication Data:

Dhar, Durga Nath.
　The chemistry of chalcones and related compounds.

　"A Wiley-Interscience publication."
　Includes index.
　1. Chalcones.　I. Title.
　QD441.D47　　547'.036　　80-39560
　ISBN 0-471-08007-1

Printed in the United States of America

10 9 8 7 6 5 4 3 2 1

Dedicated to the
Memory of
Sir Robert Robinson

Foreword

The chemical literature has grown enormously since the Second World War. It has become increasingly difficult to read meaningfully what is published. Fortunately, the literature of organic chemistry is presented in internationally readable structural formulas. It is, as a consequence, far easier to find what is novel in the literature of organic chemistry than it is in the literature of (say) biochemistry. Nevertheless, the physical burden of reading all the world's literature in organic chemistry as it becomes available is becoming a herculean task. Even if one works sixteen hours a day, one would still not manage to read carefully all that is published.

One solution to this problem, which is increasingly adopted, is not to bother to read the literature at all! One then relies on review articles, on attendance at meetings, on osmosis from one's friends and neighbors, and on specialized monographs to keep up to date.

This book is a good example of a valuable specialized monograph. It deals in detail with the chemistry of the chalcones and includes sections on physical properties and on biological activities.

However, a book like this is not only a pleasure to read as an authoritative treatise on an important subject, it also serves as a valuable reference book.

Dr. Dhar deserves the heartfelt thanks of the International Community of Organic Chemists for making available this splendid and up-to-date monograph. It will serve a very useful purpose.

SIR DEREK BARTON

Gif-sur-Yvette, France
May 1981

Preface

The chemistry of chalcone has been recognized as a significant field of study. The phenomenal growth of publications in this area is undoubtedly a reflection of the interest it is receiving throughout the world. Unfortunately no book has appeared to date dealing exclusively with the broad aspects of chalcone chemistry. Therefore, there exists a need for such a book that not only gives background information about chalcones, but also provides a bird's eye view of the entire field. The work has been extended to include some derivatives of chalcones. The literature appearing in major journals and chemical abstracts has been covered through 1979.

The general arrangement of the book is as follows. It is divided into four parts, comprising 29 chapters. Part 1 (Chapters 1–3) deals with an introduction to the subject and highlights the various methods of synthesis of substituted chalcones and includes some of the naturally occurring chalcones, such as carthamin and sophoradin. Part 2 comprises 14 chapters (Chapters 4–17), which are essentially concerned with the various reactions of chalcones, for example, their reactions with oxidizing and reducing agents, ketones, esters, amides, cyanides, amines, organometallics, and halogens. This includes cyclization (chemical as well as biochemical) and photochemical and polarographic reactions. The physical properties of chalcones, including spectroscopic, x-ray crystallographic and dipole moment measurement studies form the subject matter of Part 3 (Chapters 8–24) of the book. The color reactions, detection, and estimation of chalcones are dealt with in Chapter 20, while the chromatographic methods of separation of chalcones from other flavonoids are described in Chapter 24. A list of naturally occurring chalcones, with some of their derivatives, is given in Chapter 25. Some of the chalcones are reported to inhibit the growth of several pathogenic microorganisms and fungi and are also claimed to exhibit some interesting therapeutic properties, such as hypotensive and antipeptic ulcer activities (Chapter 26). Some of the chalcones and their derivatives have applications or are being considered for potential use, and these are covered in Chapter 29 (Part 4). Examples of these applications are artificial sweeteners, stabilizers against heat, light, and aging of several materials, analytical reagents, scintillators, polymerization catalysts, and preparation of photoconducting compositions. Chapter 28 is devoted in particular to the reactions of two important derivatives of chalcones, epoxides and α, β-dibromides. These compounds serve as useful starting materials for the preparations of a large variety of related compounds. A comprehensive subject index is included at the end of the book.

It is my privilege to thank Sir Derek Barton, D.Sc., F.R.S., Nobel Laureate, for kindly contributing the Foreword.

The writing of the concluding chapters was done during an enjoyable visit to Shemyakin Institute of Bio-organic Chemistry, U.S.S.R. Academy of Sciences, Moscow, and I take this opportunity to thank Academicians Yu. A. Ovchinnikov, and I. V. Torgov for their generous hospitality.

It is indeed a pleasure to acknowledge the assistance of my students, Drs. A. K. Banerjee, S. C. Suri, A. K. Singh, P. Dwivedi, Messrs. H. C. Misra, R. Raghunathan, and K. S. Keshavamurthy, who, by their enthusiastic cooperation and generous donation of time and effort, helped to make the book a reality. My thanks are due to Dr. S. S. Misra who undertook the arduous task of assisting me in literature survey. It is also a pleasure to record my appreciation to Messrs. A. C. Saha, Anil Kumar, and Bishember Nath for their continued cooperation.

I wish to thank the Educational Development Centre as well as the Curriculum Development Centre (Quality Improvement Programme) I.I.T. Kanpur for providing the financial assistance for the preparation of the manuscript.

Finally, I wish to express my gratitude to my wife, Rupa, and my children, Preeti and Pankaj, for their understanding and patience during the period the work was in progress.

DURGA NATH DHAR

Kanpur, India
June 1981

Contents

Abbreviations

Ac	Acetyl	DNPH	Dinitrophenylhydrazine
Alc	Alcoholic	EDTA	Ethylenediaminetetraacetic acid
Aq	Aqueous		
Ar	Aryl	Et	Ethyl
B	Base	Liq	Liquid
Bz	Benzoyl	LAH	Lithium aluminum hydride
NBS	*N*-Bromosuccinimide		
Bu	Butyl	LCP	Lithium chloropalladite
Conc	Concentrated	Me	Methyl
DME	Dimethoxyethane	NDA	*N,N*-Dimethylaniline
DMAD	Dimethylacetylene-dicarboxylate	Nu	Nucleophile
		Ph,ϕ	Phenyl
DMC	*N,N*-Dimethylchalcone	PPA	Polyphosphoric acid
DMF	Dimethylformamide	Py	Pyridine
DMSO	Dimethylsulfoxide	Satd	Saturated
DMSOM	Dimethylsulfoxonium methylide	TN	Thallic nitrate
		Ts	Tosylate

The Chemistry of
Chalcones and
Related Compounds

PART ONE
Introduction

Chapter One
Introduction

Benzylideneacetophenones constitute a class of naturally occurring pigments, which are often referred to as "chalcones." The term was first coined by Kostanecki,[1] who did pioneering work in the synthesis of natural coloring compounds. An interesting feature of chalcones (polyhydroxylated) is that they serve as starting materials for the synthesis of another class of naturally occurring and widely distributed pigments called flavones.

CLAISEN–SCHMIDT REACTION

The synthesis of chalcone, the parent member of the series, has been accomplished in a variety of ways, but perhaps the simplest method is the one involving the Claisen–Schmidt reaction. This is the reaction of acetophenone with benzaldehyde in the presence of aqueous alkali or sodium ethylate, resulting in the formation of α, β-unsaturated ketone[2]:

$$C_6H_5COCH_3 + C_6H_5CHO \xrightarrow{\text{Base}} C_6H_5COCH{:}CHC_6H_5 + H_2O$$

The substituted benzylideneacetophenones have likewise been obtained by condensing the appropriately substituted acetophenone with substituted benzaldehyde in the presence of alkali.

MECHANISM OF CHALCONE FORMATION

Kinetic studies have been reported for the base-catalyzed formation of chalcone[3-5] and its derivatives.[5,6] Two alternative mechanisms have been advanced[5] for the reaction of benzaldehyde with acetophenone in the presence of a basic catalyst:

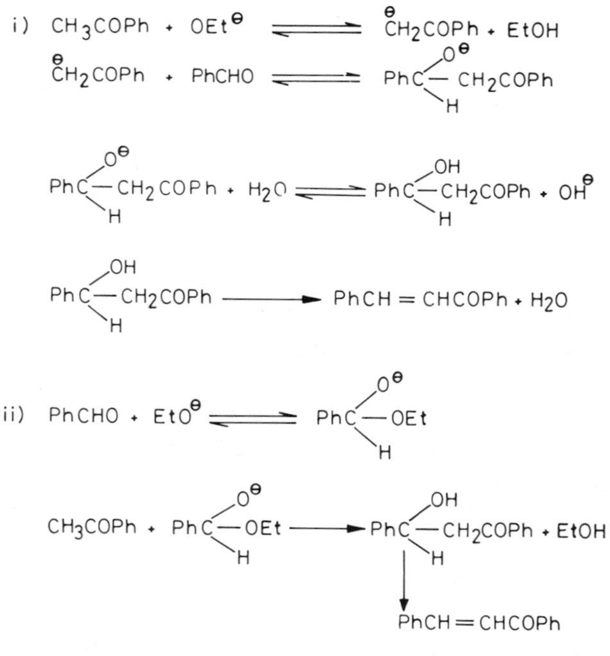

The formation of chalcone by the acid-catalyzed condensation of acetophenone and benzaldehyde has been studied.[7,8] The rate of reaction is reported to depend on the first power of the concentration of acetophenone, the first power of the concentration of benzaldehyde, and the Hammett acidity function.[7,8] Also the condensation step (see below) has been shown to be the rate-determining step in this reaction. The following mechanism seems to be operable:

$$\phi-\overset{\overset{O}{\parallel}}{C}-CH_3 \rightleftharpoons \phi-\overset{\overset{OH}{|}}{C}=CH_2$$

$$\phi-\overset{\overset{O}{\parallel}}{C}-H + \overset{\oplus}{S}H \rightleftharpoons \phi-\overset{\overset{\oplus OH}{\parallel}}{C}-H + S$$

$$(S = SOLVENT)$$

$$\phi-\overset{\overset{OH}{|}}{C}=CH_2 + \phi-\overset{\overset{\oplus OH}{\parallel}}{C}-H \longrightarrow Transition\ Complex$$

$$\longrightarrow \phi-\overset{\overset{\oplus OH}{\parallel}}{C}-CH_2-\overset{\overset{OH}{|}}{CH}-\phi$$

$$\phi-\overset{\overset{\oplus OH}{\parallel}}{C}-CH_2-\overset{\overset{OH}{|}}{CH}-\phi + S \rightleftharpoons \phi-\overset{\overset{O}{\parallel}}{C}-CH_2-\overset{\overset{OH}{|}}{CH}-\phi + \overset{\oplus}{S}H$$

$$\phi-\overset{\overset{O}{\parallel}}{C}-CH_2-\overset{\overset{\oplus OH_2}{|}}{CH}-\phi + S$$

$$\phi-\overset{\overset{O}{\parallel}}{C}-CH_2-\overset{\overset{\oplus OH_2}{|}}{CH}-\phi \longrightarrow \phi-\overset{\overset{O}{\parallel}}{C}-CH=CH-\phi + H_2O + \overset{\oplus}{H}$$

NOMENCLATURE

Benzylideneacetophenone is the parent member of the chalcone series. The substituents in the benzene rings of chalcone are numbered as shown and follow the pattern adopted by *Chemical Abstracts*.

The alternative names given to chalcone are phenyl styryl ketone, benzalacetophenone, β-phenylacrylophenone, γ-oxo-α,γ-diphenyl-α-propylene, and α-phenyl-β-benzoylethylene.[9]

REFERENCES

1   Kostanecki, S. V., and Tambor, J., *Chem. Ber.*, **32**, 1921 (1899).
2   Kostanecki, S. V., and Rossbach, G., *Chem. Ber.*, **29**, 1488 (1896).
3   Coombs, E., and Evans, D. P., *J. Chem. Soc.*, 1295 (1940).
4   Nikitin, E. K., *J. Gen. Chem. USSR* **6**, 1278 (1936).
5   Nayak, P. L. and Rout, M. K., *J. Indian Chem. Soc.*, **52**, 809 (1975).
6   Sipos, G., Fuka, A., and Széll, T., *Monatsh. Chem.*, **91**, 643 (1960).
7   Noyce, D. S. and Pryor, W. A., *J. Am. Chem. Soc.*, **77**, 1397 (1955).
8   Noyce, D. S., Pryor, W. A. and Bottini, A. H., *J. Am. Chem. Soc.*, **77**, 1402 (1955).
9   Beilstein, F., *Handbuch der Organischen Chemie*, 4th ed., Vol. 7, p. 428 (1925).

Chapter Two
Synthetic Methods

CLAISEN–SCHMIDT REACTION

A variety of methods are available for the synthesis of chalcones. The most convenient method is the one that involves the Claisen–Schmidt condensation of equimolar quantities of a substituted acetophenone with substituted benzaldehyde in the presence of aqueous alcoholic alkali.[1-36] In the Claisen–Schmidt reaction the concentration of alkali used usually ranges between 10 and 60%.[7,8,21,23,26] The reaction is carried out at about 50° for 12–15 hours or at room temperature for 1 week.[7] Under these conditions the Cannizzaro reaction[37] also takes place and thereby decreases the yield of the desired product. To avoid the disproportionation of aldehyde in the above reaction the use of benzylidene diacetate in place of aldehyde has been recommended.[38]

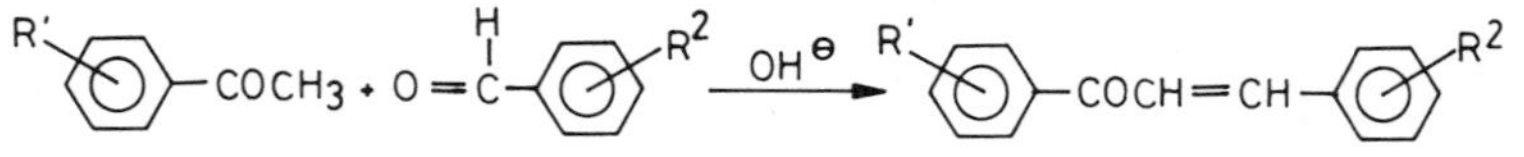

Other condensing agents have been employed, including alkali metal alkoxide,[23,25,39] magnesium *tert*-butoxide,[40] potassium carbon compound[41] (KC$_8$), hydrogen chloride,[42,43] anhydrous aluminum chloride,[44] boron trifluoride,[45] phosphorus oxychloride,[46] boric anhydride,[47] amino acids,[48] borax,[49] and organocadmium compound[50] (e.g., CdEt$_2$ in butyl ether).

In the synthesis of polyhydroxychalcones by the Claisen–Schmidt reaction a higher concentration of alkali as a condensing agent is desirable. Chalcones having a 2'-hydroxy function may cyclize to the corresponding flavanones under these conditions.[98] This difficulty has been overcome by protecting the 2'-hydroxyl group of substituted acetophenone (as methoxymethyl ether)[121] before their reaction with aromatic aldehydes. This procedure is claimed to provide a route to the synthesis of otherwise inacessible 2'-hydroxychalcones.[121]

The use of an acid catalyst, HC1, in preference to alkali, has been recommended in the synthesis of cyanomethylchalcones[42] as well as hydroxynitrochalcones.[43] 2'-Hydroxy-5-acetamidochalcones have been synthesized using phosphorus oxychloride, which is claimed to be superior to alkali as a condensing agent.[47] Condensations have been effected likewise in the presence of boric anhydride.[47] The water formed in this reaction is azeotropically distilled off with xylene. This method is reported to give good yield of chalcone. α-(Phenylthio)- and α-(phenylsulfonyl)chalcones have been prepared by the condensation of aromatic aldehydes with phenacylphenyl sulfide and phenacylsulfone, respectively.[51] The condensing agents employed in these reactions consisted of glacial acetic acid in combination with an organic base, such as piperidine or benzylamine.

Chalcone in 50% yield has also been prepared by the reaction of acetophenone and benzaldehyde under the conditions of the Perkin reaction.[122]

CHALCONE α,β-DIBROMIDES

Debromination of chalcone α,β-dibromide with 1 mole equiv of trialkylphosphine[52,54] produces chalcone in an excellent yield (92%). Triphenylphosphine[53] likewise brings about debromination of the vicinal dibromochalcone, thus

$$\phi COCHBr\text{-}CHBr\phi + \phi_3 P \longrightarrow \phi COCH = CH\phi + \phi_3 PBr_2$$

Chalcone has also been secured by the debromination of chalcone α,β-dibromide, either in the presence of chromous chloride[55] or by the action of potassium hydroxide in an acetone medium.[56]

SCHIFF'S BASES

Schiff's bases are reported to react with acetophenone and its derivative in the presence of a catalytic amount of amine hydrochloride to yield β-arylaminoketone (I).[57-59] On heating with concentrated hydrochloric acid these ketones undergo the hydramine cleavage to yield primary aromatic amine and chalcone.

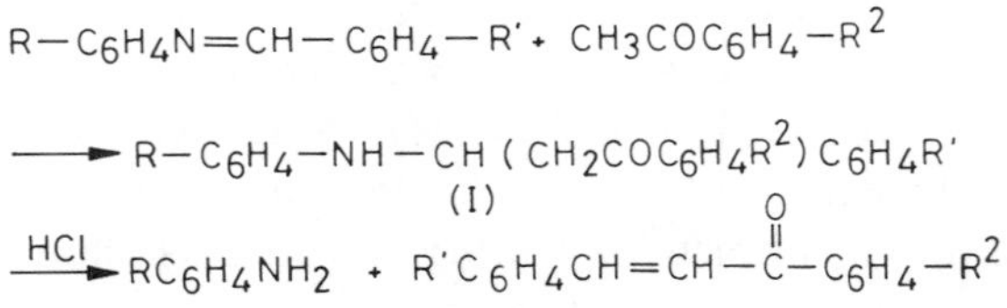

$$R-C_6H_4N=CH-C_6H_4-R' + CH_3COC_6H_4-R^2$$

$$\longrightarrow R-C_6H_4-NH-CH(CH_2COC_6H_4R^2)C_6H_4R'$$
$$\text{(I)}$$
$$\xrightarrow{HCl} RC_6H_4NH_2 + R'C_6H_4CH=CH-\overset{O}{\overset{\|}{C}}-C_6H_4-R^2$$

Hydramine cleavage is favored by the presence of electron-withdrawing substituents in β-arylamineketones.

ORGANOMETALLIC COMPOUNDS

1   Chalcone in 20% yield has been secured from acetylenic Grignard reagent[61,123] by carrying it through the following series of transformations:

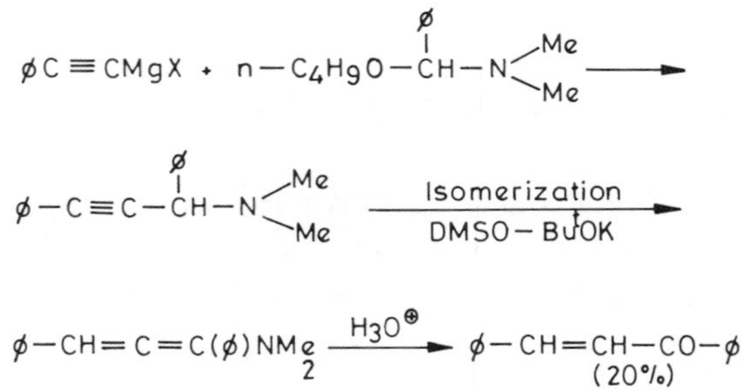

$$\phi C \equiv CMgX + n-C_4H_9O-\overset{\phi}{\overset{|}{C}}H-N\overset{Me}{\underset{Me}{<}} \longrightarrow$$

$$\phi-C\equiv C-\overset{\phi}{\overset{|}{C}}H-N\overset{Me}{\underset{Me}{<}} \xrightarrow[DMSO-Bu\overset{.}{O}K]{Isomerization}$$

$$\phi-CH=C=C(\phi)NMe_2 \xrightarrow{H_3O^\oplus} \phi-CH=CH-CO-\phi$$
$$\text{(20\%)}$$

The interaction of phenylmagnesium bromide and cinnamonitrile with ammonium chloride is reported to give an adduct, phenylstyrylketimine dimer hydrochloride.[62] The latter, on hydrolysis with dilute hydrochloric acid, furnished the chalcone.

Methylmagnesium iodide and benzaldehyde are reported to react (in the absence of ether) to give chalcone in addition to methylphenylcarbinol.[63] The formation of chalcone in this reaction arises due to the condensation of benzaldehyde with acetophenone formed by the oxidation of carbinol.

2 Chalcone has been secured (65%) by the interaction of appropriate cadmium derivative and cinnamoyl chloride in refluxing ether.[64]

3 The synthesis of chalcone (20–30%) has been achieved by the action of styryl cyanide with phenyllithium.[65] Alternatively, chalcone (33%) can be prepared by reacting *trans*-cinnamic acid with phenyllithium.[65] *cis*-Cinnamic acid is reported to react more rapidly than the *trans* isomer, producing a better yield of chalcone (57%).[65]

4 Phenylacetylene reacts with benzaldehyde at room temperature, in the presence of BF_3 in ether, to give chalcone in 60% yield.[66–68] The Lewis acid BF_3 enhances the positive nature of the carbonyl carbon, thereby facilitating the reaction.

WITTIG REACTION

Phosphorane, of the general formula $Me_nPh_{(3-n)}P{=}CHCOPh$ ($n = 0$, 1, 2, 3), is reported to react with benzaldehyde to give chalcone in good yields (70–90%).[69–71] A patent has appeared that utilizes the Wittig reaction for the preparation of chalcone in 84% yield.[72] Chalcone (60%) has also been obtained by the reaction of benzaldehyde with benzoylmethylene(*p*-dimethylaminophenyl)dimethylphosphorane[73] or phosphonate carbanion,[74] **II**, derived from diethylphenacyl phosphonate with sodium hydride:

$$(C_2H_5O)_2 \overset{\overset{O}{\uparrow}}{P} - \overset{\ominus}{C}HCO\phi \ + \ \phi - \overset{\overset{O}{\parallel}}{C} - H \xrightarrow{\text{NaH/DME}} \phi CH{=}CHCO\phi$$

$$\text{(II)}$$

Alternatively, the potassium derivative of diethylphenyl phosphonate reacts with an aromatic aldehyde (in dry toluene) to yield the desired chalcone.[75] Several substituted chalcones have been prepared by the reaction of carbonyl-stabilized phosphonium and arsonium ylides with *o*-hydroxybenzaldehydes.[76]

ENAMINES AND AROMATIC ALDEHYDES

The synthesis of chalcone has also been effected by the interaction of benzaldehyde with *N*-α-styrylmorpholine.[77]

OXIDATIVE DECARBOXYLATION OF γ-OXO ACIDS

Lead dioxide is reported to bring about the oxidative decarboxylation of 3-benzoyl-2-phenylpropionic acid to yield chalcone.[78]

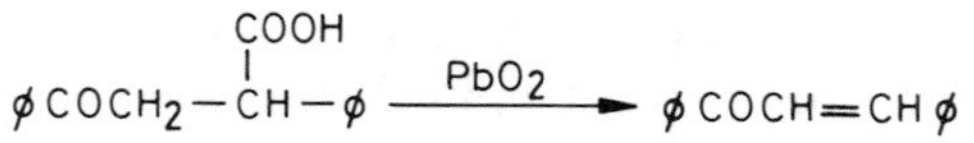

$$\phi\,COCH_2-\underset{\underset{\displaystyle CH}{|}}{\overset{\overset{\displaystyle COOH}{|}}{}}-\phi \quad \xrightarrow{\;PbO_2\;} \quad \phi\,COCH=CH\,\phi$$

PHOTO-FRIES REACTION[79–82]

Photo-Fries rearrangement of phenyl cinnamates[79,81,83] has been exploited for the synthesis of 2'-hydroxychalcone (20–50%). The same reaction has been extended to the synthesis of 2',3'-, 2',4'-, and 2',5'-dihydroxychalcones from the corresponding hydroxyphenyl cinnamates.[80] The synthesis of 2',3',5'-trihydroxychalcone has been achieved by the photolysis of 2,4-dihydroxycinnamate (the hydroxyl groups protected by methoxymethylation) followed by treatment with methanolic hydrochloric acid.[84]

BENZAL CHLORIDE AND ACETOPHENONE

Chalcone, in 75% yield, is reported to be formed by heating a mixture of benzal chloride and acetophenone at 120–130° in the presence of copper powder.[85]

METHYLBENZYLPHENACYLSULFONIUM HYDROXIDE

The synthesis of chalcone from methylbenzylphenacylsulfonium hydroxide[86] involves a number of synthetic steps, thus

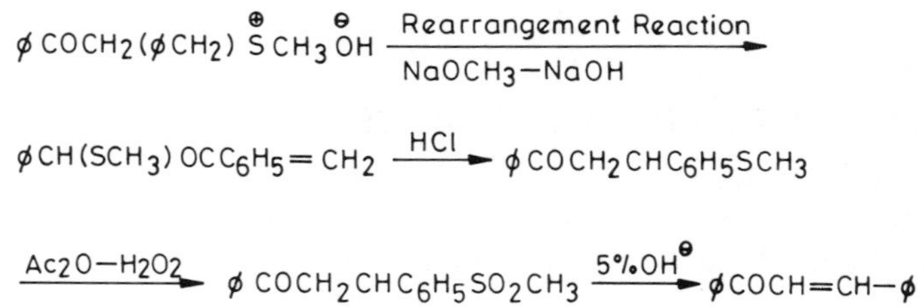

$$\phi\,COCH_2(\phi CH_2)\,\overset{\oplus}{S}\,CH_3\,\overset{\ominus}{OH} \quad \xrightarrow[\text{NaOCH}_3-\text{NaOH}]{\text{Rearrangement Reaction}}$$

$$\phi\,CH(SCH_3)\,OCC_6H_5=CH_2 \quad \xrightarrow{\;HCl\;} \quad \phi\,COCH_2CHC_6H_5SCH_3$$

$$\xrightarrow{\;Ac_2O-H_2O_2\;} \quad \phi\,COCH_2CHC_6H_5SO_2CH_3 \quad \xrightarrow{\;5\%\,OH^{\ominus}\;} \quad \phi\,COCH=CH-\phi$$

α-DIAZOACETOPHENONE

Chalcone (8%) is reported to be formed when α-diazoacetophenone is subjected to thermal decomposition.[87]

α-EPOXYCHALCONES

Chromous chloride in acetone medium brings about the reduction[88] of α-epoxychalcone to give the corresponding chalcone in a low yield.

CINNAMIC ACID ANHYDRIDE

Cadmium diphenyl[89] reacts with cinnamic acid anhydride to yield chalcone in 44% yield.

CINNAMIC ACID AND PHENOL

Polyhydroxychalcones are obtained when cinnamic acid is condensed with phenols in the presence of BF_3[90-94] or polyphosphoric acid.[95]

α-Alkylchalcones have been prepared, though in poor yield, by the reaction of 1,3,5-trihydroxyphenol with α-alkylcinnamic acid in the presence of acetic anhydride and polyphosphoric acid.[96]

CINNAMOYL CHLORIDE AND BENZENE

Chalcone (21%) along with 3-phenylhydrindone (34%) are obtained by the interaction of cinnamoyl chloride with benzene in the presence of anhydrous aluminum chloride.[97] The yield of chalcone is reported to be quantitative if chlorobenzene is also added to the reaction mixture.

CINNAMOYL CHLORIDE AND PHENOL

By using Behn's reaction polyhydroxychalcones have been synthesized.[5] It is the reaction of polyhydric phenols with cinnamoyl chloride in nitrobenzene solvent, using aluminum chloride as the condensing agent. With phloroglucinol, however, the initially formed 2',4',6'-tri-hydroxychalcone cyclizes to give 5,7-dihydroxyflavanone as the major product. It is believed that 6'-hydroxyl activates the 2'-hydroxyl to bring about the cyclization.[5]

Cinnamoyl chloride has likewise been condensed with phenolic ethers[60,98,99] in carbon disulfide to give the corresponding chalcones. The Friedel–Crafts reaction of *cis*-cinnamoyl chloride with *n*-butyl ether of thiophenol is reported to give 4'-(*n*-butylthio)chalcone.[100]

β-BENZOYLACRYLIC ACID

Aryldiazonium chlorides (carrying an electron-withdrawing substituent) react with *trans*-β-benzoylacrylic acid[101,102] to give chalcones.

$$\phi COCH=CHCOOH + ArN_2Cl \longrightarrow \phi COCH=CHAr + N_2 + CO_2 + HCl$$

The aryl group attacks the carbon atom α- of the carboxylic group, and this initial coupling is followed by decarboxylation.[101]

β-CHLOROVINYL KETONES

Substituted β-chlorovinyl ketones have been condensed with phenolic ethers[103–105] in the presence of stannic chloride to give chalcones in fairly good yield (47–65%).

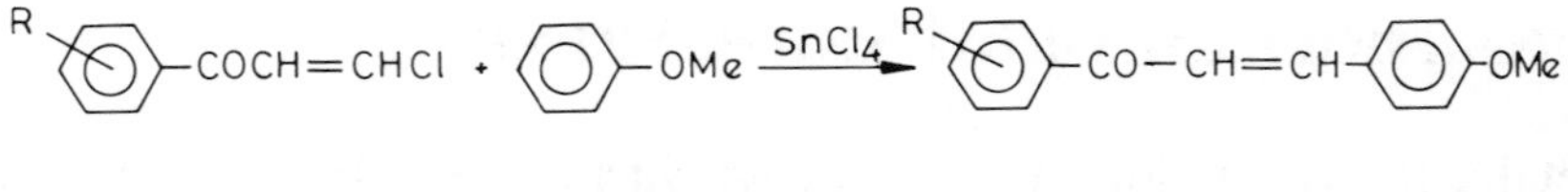

A variation of the above reaction is the interaction of β-chloro-vinyl ketone with aromatic hydrocarbons[103] and alkyl halides[103] under the influence of SnCl₄ to give the corresponding chalcones.

FLAVANONES

1 Treatment of flavanones with alkali results in the opening of γ-pyrone ring and formation of 2'-hydroxychalcone.[7,82,106–108] Thus satisfactory yield of 2'-hydroxy-4,4',6-trimethoxychalcone[7] has been obtained from 4',5,7-trimethoxyflavanone in this manner.

2 The microorganism *Gibberella fukikuroi* is capable of cleaving the oxygen heterocycle of flavanone to yield 2'-hydroxychalcone.[109] Part of the substrate undergoes ring opening, followed by microbial oxidation to yield the 2',4-dihydroxychalcone.

3 UV irradiation of flavanone is reported to yield three products, viz., 2'-hydroxychalcone (20%), 4-phenyldihydrocoumarin (13%), and salicylic acid (4%).[110]

2-DIETHYLAMINO-1, 3-DIPHENYL-2-PROPEN-1-ONE (III)

Chalcone (*cis* and *trans*) in 30% yield has been secured by prolonged irradiation of the title compound in an ethereal solution[111]:

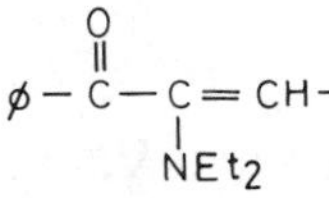

$$\phi - \overset{\overset{\displaystyle O}{\|}}{C} - \underset{\underset{\displaystyle NEt_2}{|}}{C} = CH - \phi \qquad (III)$$

MISCELLANEOUS EXAMPLES

Analogues of Carthamin

The syntheses of two analogues of carthamin, 3'-methoxy-2',4,4',6'-tetrahydroxychalcone and 2'-methoxy-3',4,4',6'-tetrahydroxychalcone, are reported in the literature (cf. Method I).[112] The hydroxylic functions in the reactants are protected by derivative preparation, methoxymethylation, and are regenerated by hydrolysis toward the end of reaction.

Sophoradin

Based on the above reaction the synthesis of sophoradin, a naturally occurring isoprenyl chalcone, has been achieved.[113]

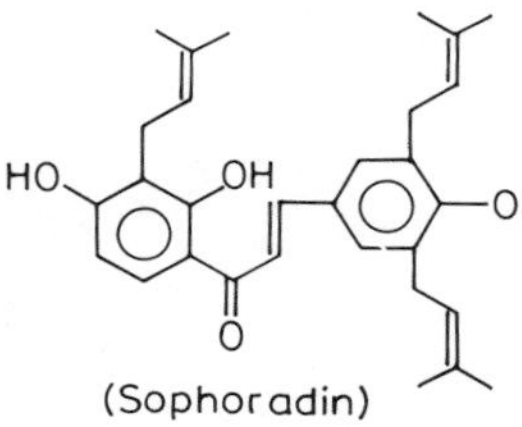

(Sophoradin)

An alternate way of sophoradin synthesis is outlined as follows[114]:

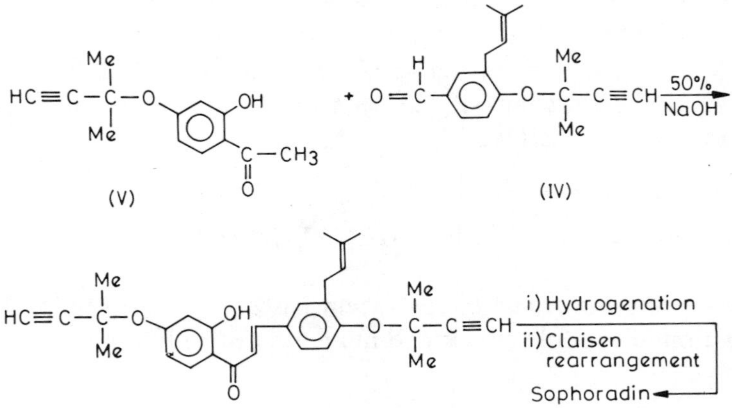

The starting materials, **IV** and **V**, are prepared as follows:

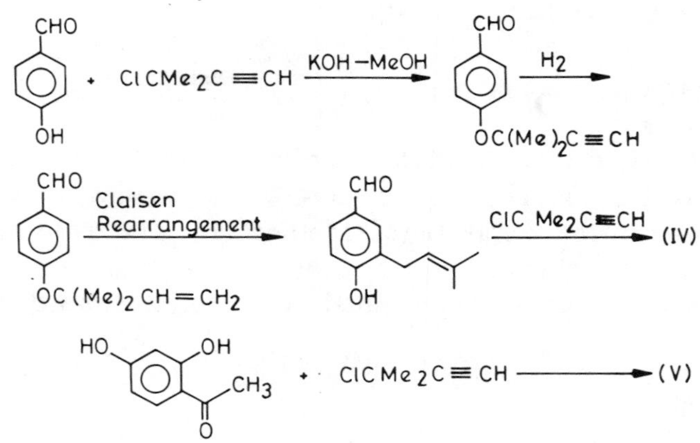

α-Aminochalcones

The synthesis of α-amino chalcones has been achieved by the following series of reactions[115]:

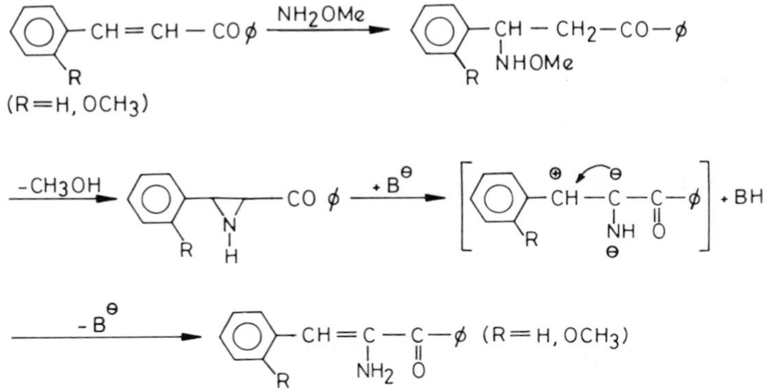

α-Hydroxychalcones

A multistep synthesis of α-hydroxychalcones has been described and involves the following sequence of reaction steps[116]:

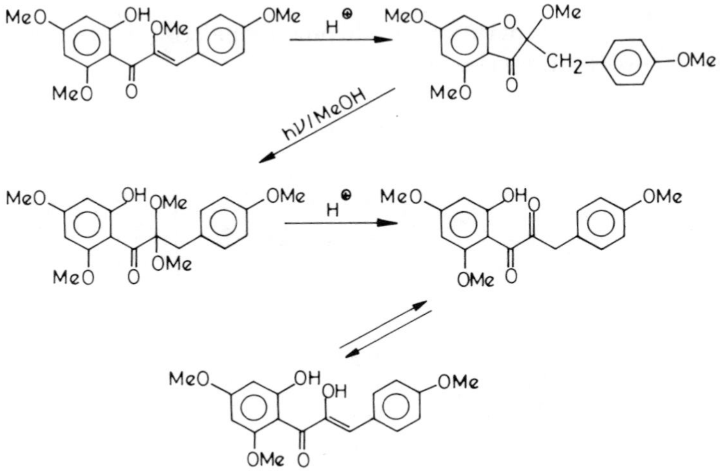

QUINOCHALCONES

The synthesis of some quinochalcones has been achieved by the oxidation of appropriately substituted hydroxychalcones.[117] The preparation of 2',4,4'-trihydroxy-3',6'-quinochalcone serves as an illustrative example:

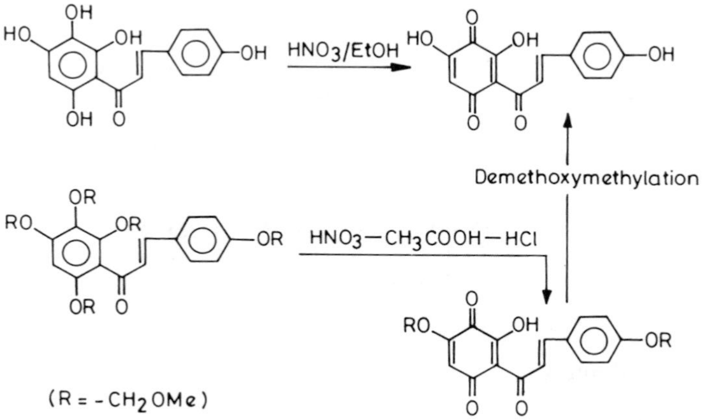

(R = -CH$_2$OMe)

CHROMENOCHALCONES

A method for the preparation of chromenochalcone is described and involves the following steps[118]:

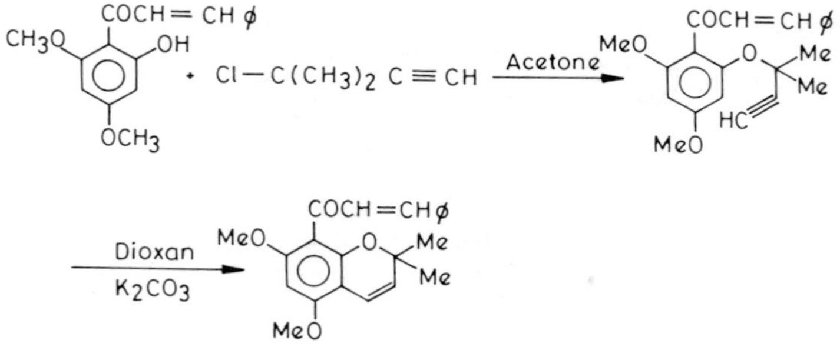

α-ARYLAZO-β-ARYLCHALCONES

Starting from α,β-diketoesters, α-arylazo-β-arylchalcones have been synthesized.[119] Thus

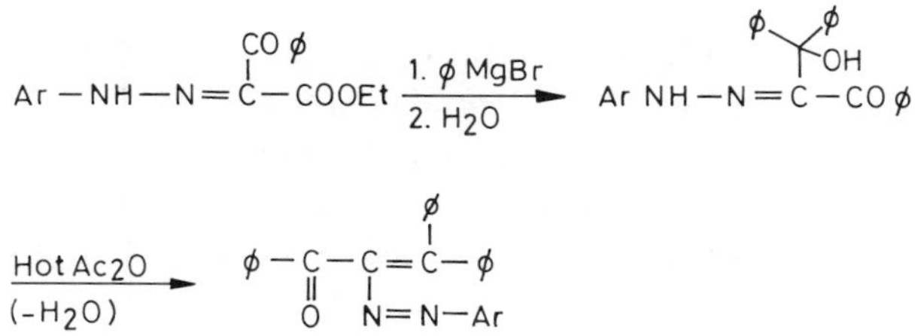

α,2'-Diacetoxy-3,4,4',6'-tetramethoxychalcone

The synthesis of α,2'-diacetoxy-3,4,4',6-tetramethoxychalcone from 2-benzyl-2-hydroxycoumaran-3-one (VI) is described.[120] Advantage is taken of the fact that the coumaranone in alkaline medium exists in equilibrium with the corresponding chalcone dianion (VII).

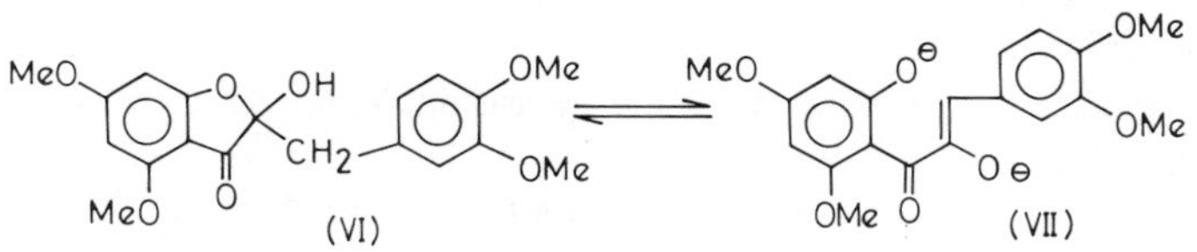

REFERENCES

1   Kohler, E. P., and Chadwell, H. M., *Org. Synth. Coll.*, **2**, 1 (1922).

2   Schraufstätter, E., and Deutsch, S., *Chem. Ber.*, **81**, 489 (1948).

2   Smith, H. E., and Paulson, M. C., *J. Am. Chem. Soc.*, **76**, 4486 (1954).

4   Obara, H., Onodera, J., and Kurihara, Y., *Bull. Chem. Soc. Japan*, **44**, 289 (1971).

5   Shinoda, J., and Sato, S., *J. Pharm. Soc. (Japan)*, **48**, 791 (1928); *Chem. Abstr.*, **23**, 836 (1929).

6   Kurth, E. F., *J. Am. Chem. Soc.*, **61**, 861 (1939).

7   Geissman, T. A., and Clinton, R. O., *J. Am. Chem. Soc.*, **68**, 697 (1946).

8   Martin, G. J., Beiler, J. M., and Avakian, S., U.S. Patent 2,769,817 (1956); *Chem. Abstr.*, **51**, 14815d (1957).

9   Jurd, L., and Horowitz, R. M., *J. Org. Chem.*, **26**, 2561 (1961).

10 Starkov, S. P., Panasenko, A. I., and Volkotrub, M. N., *Izv. Vyssh., Uched. Zaved., Khim. Khim. Tekhnol.*, **12**, 1072 (1969); *Chem. Abstr.*, **72**, 66738q (1970).

11 Vyas, G. N., and Shah, N. M., *J. Indian Chem. Soc.*, **28**, 75 (1951).

12 Matsuura, S., and Matsuura, A., *Yakugaku Zasshi*, **77**, 330 (1957); *Chem. Abstr.*, **51**, 11300i (1957).

13 Raval, A. A., and Shah, N. M., *J. Org. Chem.*, **22**, 304 (1957).

14 Shah, N. M., and Parikh, S. R., *J. Indian Chem. Soc.*, **36**, 726 (1959).

15 Dhar, D. N., *J. Indian Chem. Soc.*, **37**, 363 (1960).

16 Boichard, J., and Tirouflet, J., *Compt. Rend.*, **251**, 1394 (1960).

17 Satpathi, P. S., and Trivedi, J. P., *Curr. Sci. (India)*, **29**, 429 (1960).

18 Széll, T., *Indian J. Chem.*, **6**, 470 (1960).

19 Weber, F-G., Rinow, A., and Seedorf, C., *Z. Chem.*, **9**, 380 (1969).

20 Hsu, K. K., Lo, M. S., and Chen, F. C., *J. Chin. Chem. Soc. (Taipei)*, **16**, 91 (1969); *Chem. Abstr.*, **72**, 31393w (1970).

21 Falcao da Fonseca, L., *Bol. Fac. Farm., Univ. Coimbra, Ed. Cient*, **28**, 49 (1968); *Chem. Abstr.*, **72**, 121124p (1970).

22 Chen, F-C., and Chang, P-W., *Tai-Wan K'o Hsueh*, **24**, 105 (1970); *Chem. Abstr.*, **75**, 19859g (1971).

23 Fujise, S., and Tatsuta, H., *J. Chem. Soc. Japan*, **63**, 932 (1942).

24 Hamada, M., *Botyu-Kayaku*, **21**, 22 (1956); *Chem. Abstr.*, **51**, 3519d (1957).

25 Széll, T., and Bajusz, S., *Magy. Kém. Foly.*, **61**, 235 (1955); *Chem. Abstr.*, **52**, 9048g (1958).

26 Tsukerman, S. V., Cháng, K-S., and Lavrushin, V. F., *Zh. Obshch. Khim.*, **34**, 2881 (1964).

27 Dhar, D. N., and Singh, R. K., *J. Indian Chem. Soc.*, **48**, 83 (1971).

28 Loth, H., and Worm, G., *Arch. Pharm. (Weinheim)*, **301**, 897 (1968); *Chem. Abstr.*, **70**, 68064z (1969).

29 Biletch, H. A., and Rajunas, J. V., Jr., U.S. Patent, 3,361,827 (1968); *Chem. Abstr.*, **68**, 104756w (1968).

30 Shah, P. R., and Shah, N. M., *Curr. Sci. (India)*, **31**, 12 (1962).

31 Lespagnol, A., Lespagnol, C., Lesieur, D., Bonte, J. P., Blain, Y., and Labiau, O., *Chim. Ther.*, **6**, 192 (1971); *Chem. Abstr.*, **75**, 98285t (1971).

32 Bargellini, G., and Martegiani, E., *Gazz. Chim. Ital.*, **42**, II, 427.

33 Lavrushin, V. F., and Verkhovod, N. N., *Zh. Organ. Khim.*, **1**(7), 1220 (1965); *Chem. Abstr.*, **63**, 13118b (1965).

34 Wolf, V. L., and Tröltzsch, C., *J. Prakt. Chem.*, **17**, 69 (1962).

35 Hayakawa, G., and Inoue, T., Japanese Patent, 71 07,388 (1971); *Chem. Abstr.*, **74**, 143332y (1971).

36 Pomerantseva, L. L., Smirnova, G. A., and Oleinik, A. V., *Tr. Khim. Khim. Tekhnol.* (1), 14 (1972); *Chem. Abstr.*, **78**, 159099w (1973).

37 Dhar, D. N., and Lal, J. B., *J. Org. Chem.*, **23**, 1159 (1958).

38 Davey, W., and Gwilt, J. R., *J. Chem. Soc.*, 1008 (1957).

39 Gilman, H., and Cason, L. F., *J. Am. Chem. Soc.*, **72**, 3469 (1950).

40 Guthrie, J. L., and Rabjohn, N., *J. Org. Chem.*, **22**, 176 (1957).

41 Rochus, W., Kickuth, R., German Patent, 1,095,832 (1957); *Chem. Abstr.*, **56**, 10976f (1961).

42 Rorig, K. J., U. S. Patent 2,755, 299 (1956); *Chem. Abstr.*, **51**, 2868c (1957).

43 Sipos, G., Dobo, I., and Czukor, B., *Acta Phys. Chem. Szeged*, **8**, 160 (1962); *Chem. Abstr.*, **59**, 5059g (1963).

44 Calloway, N. O., and Green, L. D., *J. Am. Chem. Soc.*, **59**, 809 (1937).

45 Breslow, D. S., and Hauser, C. R., *J. Am. Chem. Soc.*, **62**, 2385 (1940).

46 Raval, A. A., and Shah, N. M., *J. Sci. Ind. Res. (India)*, **21B**, 234 (1962).

47 Kuskov, V. K., and Utenkova, G. N., *Zh. Obshch. Khim.*, **29**, 4030 (1959).

48 Reichel, L., *Naturwissenschaften*, **32**, 215 (1944).

49 Jadhav, G. V., and Kulkarni, V. G., *Curr. Sci. (India)*, **20**, 42 (1951).

50 Gilman, H., and Nelson, J. F., *Rec. Trav. Chim.*, **55**, 518 (1936).

51 Balaiah, V., and Natarajan, C., *Indian J. Chem.*, **8**, 694 (1970).

52 Dershowitz, S., and Proskauer, S., *J. Org. Chem.*, **26**, 3595 (1961).

53 Hoffmann, H., and Diehr, H. J., *Tetrahedron Letters*, 583 (1962).

54 Dershowitz, S., U.S. Patent, 3,104,257 (1963); *Chem. Abstr.*, **60**, P 2848f (1964).

55 Julian, P. L., Cole, W., Magnani, A., and Meyer, E. W., *J. Am. Chem. Soc.*, **67**, 1728 (1945).

56 Bickel, C. L., *J. Am. Chem. Soc.*, **72**, 349 (1950).

57 Kozlov, N. S., Pinegina, L. Y., and Selezneva, E. A., *Zh. Obshch. Khim.*, **32**, 436 (1962).

58 Kozlov, N. S., Pak, V. D., and Simonova, E. V., *Vestsi Akad. Nauk Belarus, SSR, Ser. Khim. Nauk* (4), 111 (1968); *Chem. Abstr.*, **70**, 77514y (1969).

59 Kozlov, N. S., and Shur, I. A., *Zh. Obshch. Khim.*, **29**, 2706 (1959).

60 Monti, L., *Gazz. Chim. Ital.*, **60**, 43 (1930).

61 Normant, H., and Mantione, R., *Compt. Rend.*, **259**, 1635 (1964).

62 Henze, H. R., and Swett, L. R., *J. Am. Chem. Soc.*, **73**, 4918 (1951).

63 Tarbell, D. S., and Paulson, M. C., *J. Am. Chem. Soc.*, **64**, 2842 (1942).

64 Langlais, M., Buzas, A., Soussan, G., and Freon, P., *Compt. Rend.*, **261**, 2920 (1965).

65 Bond, J. C., and Wright, G. F., *J. Am. Chem. Soc.* **72**, 1023 (1950).

66 Arens, J. F., Tjakob, H. J., and Vieregge, H., U.S. Patent 3,030,359 (1962); *Chem. Abstr.*, **57**, 11024h (1962).

67 Vieregge, H., Bos, H. J. T., and Arens, J. F., *Rec. Trav. Chim.*, **78**, 664 (1959).

68 Bos, H. J. T., and Arens, J. F., *Rec. Trav. Chim.*, **82**, 845 (1963).

69 Bestmann, H. J., and Arnason, B., *Chem. Ber.*, **95**, 1513 (1962).

70 Trippett, S., and Walker, D. M., *Chem. Ind. (London)*, 933 (1960).

71 Trippett, S., and Walker, D. M., *J. Chem. Soc.*, 1266 (1961).

72 Bestmann, H. J., and Kratzer, O., German Patent 1,256,642 (1967); *Chem. Abstr.*, **69**, 52933a (1968).

73 Trippett, S., and Walker, D. M., *J. Chem. Soc.*, 2130 (1961).

74 Wadsworth, W. S., Jr., and Emmons, W. D., *J. Am. Chem. Soc.*, **83**, 1733 (1961).

75 Grinev, G. V., Chervenyuk, G. I., and Dombrovskii, A. V., *Zh. Obshch. Khim.*, **38**, 225 (1968).

76 Bravo, P., Tricozzi, C., and Cezza, A., *Gazz. Chim. Ital.*, **105**, 109 (1975).

77 Birkofer, L., Kim, S. M., and Engels, H. D., *Chem. Ber.*, **95**, 1495 (1962).

78 Hertzler, D. V., Berdahl, J. M., and Eisenbraun, E. J., *J. Org. Chem.*, **33**, 2008 (1968).

79 Ramakrishnan, V. T., and Kagan, J., *J. Org. Chem.*, **35**, 2901 (1970).

80 Obara, H., Takahashi, H., and Hirano, H., *Bull. Chem. Soc. Japan*, **42**, 560 (1969).

81 Obara, H., Takahashi, H., and Onodera, J., *Kogyo Kagaku Zasshi*, **72**, 309 (1969); *Chem. Abstr.*, **71**, 12758g (1969).

82 Wilson, C. W., U.S. Patent 2,425,291 (1947); *Chem. Abstr.*, **41**, 6674c (1947).

83 Bhatia, V. K., and Kagan, J., *Chem. Ind. (London)*, 1203 (1970).

84 Onodera, J., and Obara, H., *Bull. Chem. Soc. (Japan)*, **47**, 240 (1974).

85 Tsukervanik, I. P., and Galust'yan, G. G., *Zh. Obshch. Khim.*, **31**, 528 (1961).

86 Ruiz, E. B., *Acta Salmenticensia, Ser. Cienc.*, **2**, No. 7, 64 (1958); *Chem. Abstr.*, **54**, 7624b (1960).

87 Yates, P., and Clark, T. J., *Tetrahedron Letters*, 435 (1961).

88 Julian, P. L., and Cole, J. W., U.S. Patent 3,055,918 (1962); *Chem. Abstr.*, **58**, P 5767d (1963).

89 Nightingale, D., and Wadsworth, F., *J. Am. Chem. Soc.*, **67**, 416 (1943).

90 Mani, R., and Venkataraman, K., *Curr. Sci. (India)*, **23**, 220 (1954).

91 Starkov, S. P., and Panasenko, A. I., *Zh. Organ. Khim.*, **7**, 1463 (1971); *Chem. Abstr.*, **75**, 129462e (1971).

92 Starkov, S. P., Panasenko, A. I., and Volkotrub, M. N., *Izv. Vyssh. Ucheb. Zaved., Khim. Khim. Tekhnol.*, **15**, 866 (1972); *Chem. Abstr.*, **77**, 114191h (1972).

93 Starkov, S. P., Panasenko, A. I., and Volkotrub, M. N., *Izv. Vyssh. Ucheb. Zaved., Khim. Khim. Tekhnol.*, **16**, 895 (1973); *Chem. Abstr.*, **79**, 91907w (1973).

94 Starkov, S. P., Starkova, S. K., and Goncharenko, G. A., *Izv. Vyssh. Ucheb. Zaved., Khim. Khim. Tekhnol.*, **20**, 1149 (1977); *Chem. Abstr.*, **88**, 22272j (1978).

95 Reichel, L., and Proksch, G., *Justus Liebigs Ann. Chem.*, **745**, 59 (1971).

96 Hasebe, N., *Nippon Kagaku Zasshi*, **89**, 534 (1968); *Chem. Abstr.*, **70**, 3455h (1969).

97 Rasschaert, A., Janssens, W., and Slootmaekers, P. J., *Bull. Soc. Chim. Belges.*, **75**, 449 (1966); *Chem. Abstr.*, **66**, 2305e (1967).

98 Bodani, D. C., Bodani, V. V., and Wheeler, T. S., *Curr. Sci. (India)*, **6**, 604 (1938).

99 Bargellini, G., and Finkelstein, M., *Gazz. Chim. Ital.*, **42**, II, 417.

100 Velarde, E., Hammond, G., and Hoet, P., *Bol. Soc. Quim. Peru*, **41**, 133 (1975); *Chem. Abstr.*, **85**, 123504e (1976).

101 Mehra, H. S., and Mathur, K. B. L., *J. Indian Chem. Soc.*, **32**, 465 (1955).

102 Mehra, H. S., and Mathur, K. B. L., *J. Indian Chem. Soc.*, **33**, 618 (1956).

103 Kochetkov, N. K., and Belyaev, V. F., *Zh. Obshch. Khim.*, **30**, 1495 (1960).

104 Belyaev, V. F., Yatsevich, N. M., and Sokolov, N. A., *Zh. Obshch. Khim.*, **32**, 2022 (1962).

105 Belyaev, V. F., *Zh. Obshch. Khim.*, **34**, 861 (1964).

106 Bargellini, G., and Monti, L., *Gazz. Chim. Ital.*, **44**, II, 24 (1914).

107 Linke, H. A. B., and Eveleigh, D. E., *Z. Naturforsch.*, **30B**, 606 (1975).

108 Linke, H. A. B., and Eveleigh, D. E., *Z. Naturforsch.*, **30B**, 940 (1975).

109 Udupa, S. R., Banerji, A., and Chadha, M. S., *Tetrahedron*, **25**, 5415 (1969).

110 Mack, P. O. L., and Pinhey, J. T., *Chem. Commun.* (8), 451 (1972).

111 Arnould, J. C., and Pete, J. P., *Tetrahedron Letters*, 2459 (1975).

112 Obara, H., Onodera, J., and Abe, S., *Chem. Letters*, 335 (1974).

113 Kyogoku, K., Hatayama, K., Yokomori, S., Seki, T., and Tanaka, I., *Agric. Biol. Chem.*, **39**, 133 (1975); *Chem. Abstr.*, **82**, 156524c (1975).

114 Kyogoku, K., Hatayama, K., Yokomori, S., Seki, T., and Tanaka, I., *Agric. Biol. Chem.*, **38**, 2291 (1974); *Chem. Abstr.*, **82**, 97789t (1975).

115 Reichel, L., and Pritze, P., *Justus Liebigs Ann. Chem.*, 120 (1974).

116 Fourie, T. G., Ferreira, D., and Roux, D. G., *Chem. Commun.*, 760 (1974).

117 Obara, H., and Onodera, J., *Chem. Letters*, 1357 (1974).

118 Mathur, A. K., Mathur, K. B. L., and Seshadri, T. R., *Indian J. Chem.*, **11**, 1231 (1973).

119 Harhash, A. H., Mansour, A. K., Elnagdi, M. H., and Elmoghayar, M. R. H., *J. Prakt. Chem.*, **315**, 235 (1973).

120 Clark-Lewis, J. W., and Jemison, R. W., *Aust. J. Chem.*, **21**, 815 (1968).

121 Rall, G. J. H., Overholzer, M. E., Ferreira, D., and Roux, D. G., *Tetrahedron Letters*, 1033 (1976).

122 Staněk, J., and Horák, M. *Coll. Czech. Chem. Commun.*, **15**, 1037 (1951).

123 Normant, H. (Société des Usines Chimiques Rhone-Poulene) French Patent 1,411,091 (1965); *Chem. Abstr.*, **64**, 3426e (1966).

Chapter Three

Some Classes of Chalcones and Chalcone Analogues

SUBSTITUTED CHALCONES

Hydroxynitrochalcones

Several hydroxynitrochalcones[1-11] have been prepared using Claisen–Schmidt condensation. This condensation has been effected with a variety of condensing agents, viz., aqueous alkali,[12-17] sodium methylate, piperidine,[14] aluminum chloride,[12,14,18] boron trifluoride,[14] and dry hydrogen chloride.[14-16] The preparation of α-methylchalcones (containing a nitro group) by the interaction of hydroxynitropropiophenone with various aldehydes is also reported in the literature.[19] It has been suggested that dypnones may be the possible intermediates in the synthesis of nitrohydroxychalcones.[20]

In general the electron-donating substituents[16] (low Hammett constant) in the aldehydic component and the electron-withdrawing substituents (high Hammett constant) in the ketone[21] favor Claisen–Schmidt condensation by HCl. On the other hand, the electron-withdrawing substituent[15] in the aldehydic component favors condensation by caustic alkali.

The preparation of a large number of substituted chalcones, which include hydroxy-, methoxy-, and carboxychalcones, are reported in the literature.[22–48]

Hydroxychalcones

Some of the difficultly preparable 2'-hydroxychalcones have been synthesized in excellent yields by converting the 2-hydroxy group of 2-hydroxyacetophenone to corresponding methoxymethyl ether[49] prior to Claisen–Schmidt condensation. The hydroxychalcone is then regenerated by treatment with hot acid.

The synthesis of chalcones derived from 2-hydroxyacetophenone,[50] 2'-hydroxy-4-ethoxyacetophenone,[51] 2-hydroxy-4-n-butoxyacetophenone,[52] and 2-hydroxy-4-n-propoxy-5-bromo- (or nitro-) acetophenone[53–55] has also been reported. The preparation of polyhydroxychalcones derived from resacetophenone[32,56] and quinacetophenone[57] has been secured likewise.

The reaction of p-hydroxybenzaldehyde with substituted acetophenone is claimed to proceed better in acid than in alkaline medium.[58] This observation has been rationalized as follows: In alkaline medium, reaction i is favored, which obviously lowers the reactivity of the carbonyl carbon and hence does not lead to chalcone formation[58]:

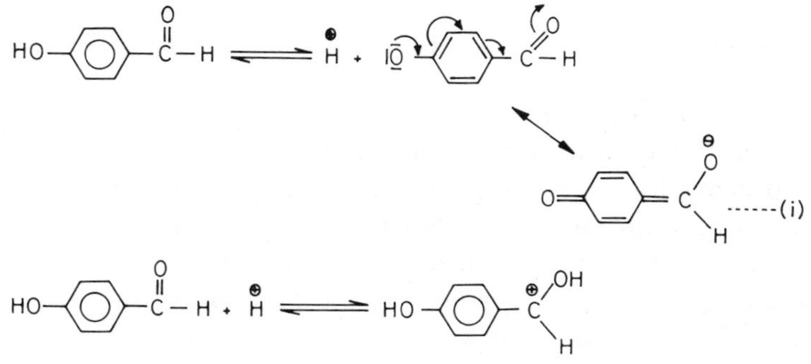

In acid medium, however, the dissociation of the phenol is restricted, and the formation of carbonium ion (I) is facilitated. These effects therefore exert a favorable influence on chalcone formation.

Methoxychalcones

The syntheses of a large number of methoxychalcones,[59,60] including those derived from quinacetophenone monomethyl ether[61] and resacetophenone dimethyl ether,[56] are described in the literature.

Carboxychalcones and Chalcone Analogues

Carboxychalcones,[62,63] including the chalcone derived from caffeic acid,[64] have been prepared by the usual Claisen–Schmidt reaction. In general the ketones required in the reaction are derived by the Fries migration of o- and p-acetoxybenzoic acids.

The naphthalene,[65-72] phenanthrene,[66] anthracene,[66] and bicyclic[73] analogues of chalcones have also been prepared.

Isoprenylchalcones

The synthesis of several naturally occurring isoprenylchalcones[74,75] has been accomplished by employing the Claisen–Schmidt reaction, viz., cordoin,[75] isocordoin,[75] 4-hydroxycordoin,[75] derricin,[75,77] 4-hydroxyderricin,[75] sophoradin,[76] and derricidin.[77]

Phosphorylated and Sulfuric Acid Esters of Chalcones

The preparation of the title compounds[78,79] has been achieved by the Claisen–Schmidt reaction.

Phenylchalcones

The synthesis of 4- (and 4'-) phenylchalcones[80] and α- (and β-) phenylchalcones[81-83] is described in the literature.

The preparation of α-phenylchalcone[82] has been secured by the interaction of appropriately substituted acrylyl chloride with phenol in the presence of aluminum chloride. On the other hand, 4'-nitro-α-phenylchalcone has been obtained by the reaction of 4'-nitrodesoxybenzoin with ethyl benzylideneacetoacetate under the influence of ethanolic sodium methoxide.[83]

Polyfluorochalcones

The polyfluorochalcones (**II–IV**) have been synthesized by the Claisen–Schmidt reaction[84]:

$$R—CH{=}CHCOR'$$

II	$R = C_6F_5$;	$R' = C_6H_5$
III	$R = C_6H_5$;	$R' = C_6F_5$
IV	$R = C_6F_5$;	$R' = C_6F_5$

In the synthesis of **IV** a lower concentration of alkali (1.5%) has been recommended.[84] With higher concentration of alkali the initially formed polyfluorochalcone undergoes a haloform type of cleavage,[84] leading to the formation of 2,3,4,5,6-pentafluoro-*trans*-cinnamic acid and pentafluorobenzene.

HETEROCYCLIC CHALCONE ANALOGUES

A number of heterocyclic analogues of chalcones have been reported. These include the chalcone analogues of pyrrole,[85,87] indole,[88,89] carbazole,[90] furan,[67,85,86,91–95] thiophene,[66,85,92,96–103] selenophene,[104] pyridine,[66,92,93,98,105–107] quinoline,[66,93] 8-hydroxyquinoline,[108] piperidine,[109] 1,4-benzodioxans,[110] pyridone,[111] pyrimidine,[112] pyrazole,[113] and acridine.[114]

Several chalcone analogues have been prepared by the Claisen–Schmidt reaction: ferrocene-,[115,116] cymantrene-,[117] benzochrotrene-,[117] cinnamoyl-,[118] chromono-,[122] chromeno-,[123,124] and coumarinochalcones.[119]

The syntheses of chromenochalcone,[120] including Flemi–Chapparin A, have been achieved (53–63%) by the reaction of 6-acetyl-5,8-dihydroxy-2,2-dimethylchromene (**V**) and substituted benzaldehydes:

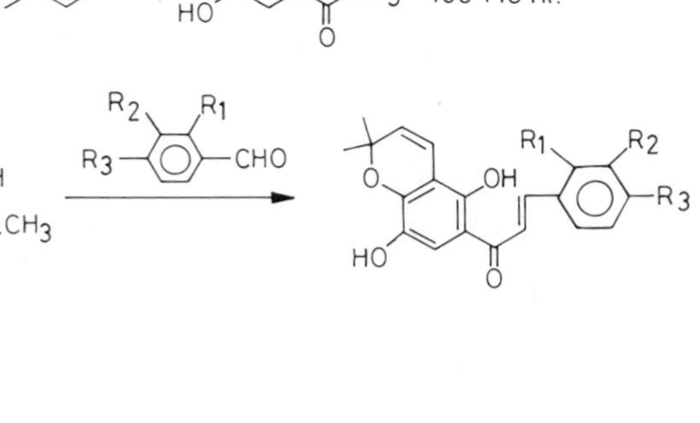

The preparation of some α-aroylchalcones involving the Knoevenagel reaction between diaroylmethanes and aromatic aldehydes has also been described.[121]

REFERENCES

1 Tanasescu, I., and Baciu, A., *Bull. Soc. Chim.* [5], **4**, 1742 (1937); *Chem. Abstr.*, **32**, 1674 (1938).

2 Kulkarni, V. G., and Jadhav, G. V., *J. Univ. Bombay*, **22A**, Part 5, No. 35, 17 (1954); *Chem. Abstr.*, **49**, 11590c (1955).

3 Széll, T., and Bajusz, S., *Magy. Kem. Foly.*, **60**, 5 (1954); *Chem. Abstr.*, **52**, 5352g (1958).

4 Chhaya, G. S., Trivedi, P. L., and Jadhav, G. V., *J. Univ. Bombay*, **26A**, Part 3, 16 (1957); *Chem. Abstr.*, **53**, 10125a (1959).

5 Chhaya, G. S., Trivedi, P. L., and Jadav, G. V., *J. Univ. Bombay*, **27A**, Part 3, 26 (1958); *Chem. Abstr.*, **54**, 8807e (1960).

6 Seshadri, S., and Trivedi, P. L., *J. Org. Chem.*, **22**, 1633 (1957).

7 Dhar, D. N., *J. Org. Chem.*, **25**, 1247 (1960).

8 Price, D., and Bogert, M. T., *J. Am. Chem. Soc.*, **56**, 2442 (1934).

9 Széll, T., *Magy. Kem. Foly.*, **64**, 44 (1958); *Chem. Abstr.*, **52**, 14577c (1958).

10 Széll, T., *Chem. Ber.*, **93**, 1928 (1960).

11 Wagh, S. P., and Jadhav, G. V., *J. Univ. Bombay*, **27A**, Part 3, 1 (1958); *Chem. Abstr.*, **54**, 8723b (1960).

12 Széll, T., *Chem. Ber.*, **92**, 1672 (1959).

13 Sipos, G., Széll, T., and Va'rnai, I., *Acta Phys. Chem. Szeged.*, **6**, 109 (1960); *Chem. Abstr.*, **55**, 14377c (1961).

14 Széll, T., and Sipos, G., *Annalen*, **641**, 113 (1961).

15 Széll, T., and Sohár, I., *Can. J. Chem.*, **47**, 1254 (1969).

16 Széll, T., and Sohár, I., *Acta Chim. (Budapest)*, **62**, 429 (1969).

17 Széll, T., Dudas, T., and Zarandy, M. S., *Acta Phys. Chem. Szeged.*, **15**, 157 (1969); *Chem. Abstr.*, **73**, 3618a (1970).

18 Sipos, G., and Széll, T., *Acta Univ. Acta Phys. Chem. Szeged.*, **5**, 70 (1959); *Chem. Abstr.*, **54**, 10951d (1960).

19 Széll, T., *J. Prakt. Chem.*, **17**, 346 (1962).

20 Sipos, G., and Széll, T., *Naturwissenschaften*, **46**, 532 (1959).

21 Sipos, G., Cseh, I., and Kalmar, A., *Acta Phys. Chem. Szeged.*, **16**, 45 (1970); *Chem. Abstr.*, **73**, 119961m (1970).

22 Kulkarni, V. G., and Jadhav, G. V., *J. Indian Chem. Soc.*, **31**, 746 (1954).

23 Kulkarni, V. G., and Jadhav, G. V., *J. Indian Chem. Soc.*, **32**, 97 (1955).

24 Patel, C. C., and Shah, N. M., *J. Indian Chem. Soc.*, **31**, 867 (1954).

25 Atchabba, F. A., Trivedi, P. L., and Jadhav, G. V., *J. Indian Chem. Soc.*, **32**, 206 (1955).

26 Davey, W., and Gwilt, J. R., *J. Chem. Soc.*, 1008 (1957).

27 Kulkarni, V. G., and Jadhav, G. V., *J. Univ. Bombay*, **23A**, 14 (1965); *Chem. Abstr.*, **51**, 11301d (1957).

28 Raval, A. A., and Shah, N. M., *J. Org. Chem.*, **21**, 1408 (1956).

29 Buu-Hoi, N. P., Sy, M., and Riche, J., *J. Org. Chem.*, **22**, 668 (1957).

30 Chhaya, G. S., Trivedi, P. L. and Jadhav, G. V., *J. Univ. Bombay*, **25**, 8 (1957); *Chem. Abstr.*, **52**, 14598h (1958).

31 Atchaba, F. A., Trivedi, P. L., and Jadhav, G. V., *J. Univ. Bombay*, **25**, 1 (1957); *Chem. Abstr.*, **52**, 14599i (1958).

32 Dhar, D. N., and Lal, J. B., *J. Org. Chem.*, **23**, 1159 (1958).

33 Széll, T., *Chem. Ber.*, **91**, 2609 (1958).

34 Chhaya, G. S., Trivedi, P. L., and Jadhav, G. V., *J. Univ. Bombay*, **26A**, 22 (1958); *Chem. Abstr.*, **53**, 14038h (1959).

35 Wagh, S. P., and Jadhav, G. V., *J. Univ. Bombay*, **26A**, 28 (1958); *Chem. Abstr.*, **53**, 14097c (1959).

36 Wagh, S. P., and Jadhav, G. V., *J. Univ. Bombay*, **26A**, 4 (1958); *Chem. Abstr.*, **53**, 14098b (1959).

37 Raut, K. B., *Diss. Abstr.*, **20**, 97 (1959).

38 Atchabba, F. A., Trivedi, P. L., and Jadhav, G. V., *J. Univ. Bombay*, **27A**, 8 (1958); *Chem. Abstr.*, **54**, 9907e (1960).

39 Mulchandani, N. B., and Shah, N. M., *Chem. Ber.*, **93**, 1918 (1960).

40 Awad, W. I., Neweihy, M. F. E., and Selim, S. F., *J. Org. Chem.*, **25**, 1333 (1960).

41 Klinke, P., and Gibian, H., *Chem. Ber.*, **94**, 26 (1961).

42 Dhar, D. N., *J. Indian Chem. Soc.*, **37**, 799 (1960).

43 Dhar, D. N., *J. Indian Chem. Soc.*, **38**, 823 (1961).

44 Merchant, J. R., and Choughuley, A. S. U., *Chem. Ber.*, **95**, 1792 (1962).

45 Seikel, M. K., Lounsbury, M. J. and Wang, S. C., *J. Org. Chem.*, **27**, 2952 (1962).

46 Jacob, K. C., Jadhav, G. V., and Vakharia, M. N., *Pesticides*, **6**, 94 (1972); *Chem. Abstr.*, **78**, 147867g (1973).

47 Shah, P. R., and Shah, N. M., *Chem. Ber.*, **97**, 1453 (1964).

48 Wagner, G., Griessmann, R., and Ritchter, P., *Pharmazie*, **34**, 65 (1979).

49 Rall, G. J. H., Oberholzer, M. E., Ferreira, D., and Roux, D. G., *Tetrahedron Letters*, 1033 (1976).

50 Dubnitskaya, V. I., Oganesyan, E. T., Shinkarenko, A. L., *Izv. Sev. Kavk. Nauchn. Tsentra Vyssh Shk., Ser. Estestv. Nauk*, **3**, 74 (1975); *Chem. Abstr.*, **83**, 205884b (1975).

51 Joshi, R. S., and Naik, H. B., *J. Inst. Chem. (India)*, **50**, 153 (1978).

52 Mankiwala, S. C., Naik, H. B., and Thakor, V. M., *J. Indian Chem. Soc.*, **52**, 560 (1975).

53 Mankiwala, S. C., Naik, H. B., and Thakor, V. M., *J. Inst. Chem. (India)*, **47**, 132 (1975).

54 Shah, N. J., Jhaveri, L. C., and Naik, H. B., *J. Inst. Chem. (India)*, **49**, 204 (1977).

55 Shah, N. J., Jhaveri, L. C., and Naik, H. B., *J. Inst. Chem. (India)*, **50**, 171 (1978).

56 Lal, J. B., *J. Indian chem. Soc.*, **16**, 296 (1939).

57 Vyas, G. N., and Shah, N. M., *J. Indian Chem. Soc.*, **26**, 273 (1949).

58 Sipos, G., and Sirokman, F., *Nature*, 202 (4931), 489 (1964).

59 Kuroda, C., and Matsukuma, T., *Sci. Papers Inst. Phys. Chem. Res. (Tokyo)*, **18**, 51 (1932); *Chem. Abstr.*, **26**, 2442 (1932).

60 Kuroda, C., and Nakamura, T., *Sci. Papers Inst. Phys. Chem. Res. (Tokyo)*, **18**, 61 (1932); *Chem. Abstr.*, **26**, 2442 (1932).

61 Vyas, G. N., and Shah, N. M., *Curr. Sci. (India)*, **19**, 318 (1950).

62 Shah, D. N., and Shah, N. M., *J. Indian Chem. Soc.*, **26**, 235 (1949).

63 Vyas, G. N., and Shah, N. M., *J. Indian Chem. Soc.*, **28**, 41 (1951).

64 Cahn, J., and Wermuth, C. G., French Patent 1,583,930; *Chem. Abstr.*, **73**, 76884u (1970).

65 Belyaev, V. F., Abrazhevich, A. I., and Vasilevich, I. F., *Vestsi Akad. Nauk Belarus SSR, Ser. Khim. Nauk* (1), 116 (1971); *Chem. Abstr.*, **74**, 125243r (1971).

66 Raut, K. B., and Wender, S. H., *J. Org. Chem.*, **25**, 50 (1960).

67 Tewari, S. S., and Singh, A., *J. Indian Chem. Soc.*, **38**, 931 (1961).

68 Misra, S. S., and Dinkar, *J. Indian Chem. Soc.*, **49**, 725 (1972).

69 Wagh, S. P., and Jadhav, G. V., *J. Univ. Bombay*, **25A**, 23 (1957); *Chem. Abstr.*, **52**, 7304h (1958).

70 Deshmukh, G. V., and Wheeler, T. S., *J. Chem. Soc.*, 96 (1939).

71 Mitina, V. G., Zadorozhnyi, B. A., *Vestn. Kharkov un-ta Khim.*, **115**, 108 (1974); *Chem. Abstr.*, **83**, 114046p (1975).

72 Yuldashev, K. Y., *Dokl. Akad. Nauk Uzb. SSR* (12), 35 (1977); *Chem. Abstr.*, **91**, 74372y (1979).

73 Raikova, T. S., *Vestsi Akad. Nauk Belarus, SSR, Ser. Khim. Nauk,* (3), 128 (1973); *Chem. Abstr.*, **79**, 65949s (1973).

74 Kyogoku, K., Hatayama, K., Yokomori, S., and Seki, T., Japanese Patent 74,126,654 (1974); *Chem. Abstr.*, **83**, 78864t (1975).

75 Lupi, A., Delle Monache, G., Delle Monache, F., Marini-Bettolo, G. B., Goncalves de Lima, O., and De Mello, J. F., *Farmaco, Ed. Sci.*, **30**, 449 (1975); *Chem. Abstr.*, **83**, 178450y (1975).

76 Kyogoku, K., Hatayama, K., Yokomori, S., Seki, T., and Tanaka, I., *Agric. Biol. Chem.*, **39**, 667 (1975); *Chem. Abstr.*, **83**, 28382p (1975).

77 Khanna, R. N., Khanna, P. L., Manchanda, V. P., and Seshadri, T. R., *Indian J. Chem.*, **11**, 1225 (1973).

78 Bergl, F., Cohen, A., Haworth, J. W., and Hughes, E. G., U.S. Patent, 2,465,320 (1949); *Chem. Abstr.*, **43**, 6235h (1949).

79 Yamaguchi, K., *Nippon Kagaku Zasshi*, **84**, 148 (1963).

80 Hsu, K. K., Wu, T. S., and Shi, J. Y., *J. Chin. Chem. Soc. (Taipei)*, **19**, 45 (1972); *Chem. Abstr.*, **77**, 74967g (1972).

81 Offe, H. A., *Chem. Ber.*, **80**, 449 (1947).

82 Vorländer, D., Osterburg, J., and Meye, O., *Chem. Ber.*, **56B**, 1136 (1923).

83 Buza, D., and Gryff-Keller, A., *Rocz. Chem.*, **43**, 1945 (1969); *Chem. Abstr.*, **72**, 89628t (1970).

84 Filler, R., Beaucaire, V. D., and Kang, H. H., *J. Org. Chem.*, **40**, 935 (1975).

85 Corvaisier, A., *Bull. Soc. Chim. France*, 528 (1962).

86 Tsukerman, S. V., Izvekov, V. P., and Lavrushin, V. F., *Khim. Geterotsikl. Soedin, Akad. Nauk. Latv. SSR*, 387 (1966); *Chem. Abstr.*, **65**, 12159a (1966).

87 Dhar, D. N., and Singh, R. K., *Zh. Obshch. Khim.*, **40**, 1156 (1970).

88 Venturella, P., Bellino, A., and Piozzi, F., *Farmco, Ed. Sci.*, **26**, 591 (1971); *Chem. Abstr.*, **75**, 88432z (1971).

89 Stepanova, G. P., Sakharova, N. A., and Stepanov, B. I., *Izv. Vyssh. Ucheb. Zaved., Khim. Khim. Tekhnol.*, **14**, 83 (1971); *Chem. Abstr.*, **75**, 35593w (1971).

90 Belyaev, V. F., Grushevich, V. I., and Abrazhevich, A. I., *Zh. Organ. Khim.*, **7**, 610 (1971); *Chem. Abstr.*, **75**, 5617q (1971).

91 Deutsch, D. H., and Garcia, E. N., U.S. Patent 2,754,299 (1956); *Chem. Abstr.*, **51**, 4437h (1957).

92 Buu-Hoi, N. P., Xuong, N. D., and Trieu, T. C., *Bull. Soc. Chim. France*, 584 (1961); *Chem. Abstr.*, **56**, 5907f (1962).

93 Ariyan, Z. S., and Suschitzky, H., *J. Chem. Soc.*, 2242 (1961).

94 Velarde, E., *Bol. Soc. Quim. Peru*, **36**, 127 (1970); *Chem. Abstr.*, **75**, 19808q (1971).

95 Jurasek, A., Knoppova, V., Danderova, M., Kovac, J., and Reinprecht, L., *Tetrahedron*, **34**, 1833 (1978).

96 Churkin, Y. D., and Savin, V. I., *Khim. Geterotsikl. Soedin.*, Sb. No. 3, (1971) (Serusoder-Zhashchie Geterotsikly), 60; *Chem. Abstr.*, **77**, 88189s (1972).

97 Hanson, G. A., *Bull. Soc. Chim. Belges*, **67**, 712 (1958); *Chem. Abstr.*, **53**, 17995c (1959).

98 Devitt, P. F., Timoney, A., and Vickars, M. A., *J. Org. Chem.*, **26**, 4941 (1961).

99 Belyaev, V. F., and Abrazhevich, A. I., *Khim. Geterotsikl. Soedin.*, 228 (1967); *Chem. Abstr.*, **67**, 73466z (1967).

100 Belyaev, V. F., and Abrazhevich, A. I., *Khim. Geterotsikl. Soedin.*, 827 (1967); *Chem. Abstr.*, **68**, 114331t (1968).

101 Real, L., Jone, M., Hilda, W., and George, M., *Can. J. Chem.*, **46**, 1952 (1968).

102 Verkhovod, V. M., Roberman, A. I., Ostrovskaya, B. I., Verkhovod, N. N., and Lavrushin, V. F., *Vestn. Kharkov un-ta*, 113 (1974) (115 Khimya Vyp. 5); from *Ref. Zh., Khim.* 1975; Abstr., No. 1Zh249; *Chem. Abstr.*, **83**, 147361r (1975).

103   Fisera, L., Kovac, J., and Hrabovsky, J., *Pr. Chem. Fak.* SVST 1975–76 (Pub. 1978); *Chem. Abstr.*, **90**, 137601z (1979).

104   Tsukerman, S. V., Orlov, V. D., Izvekov, V. P., Lavrushin, V. F., and Yur'ev, Y. K., *Khim. Geterotsikl. Soedin., Akad. Nauk Latv. SSR* (1), 34 (1966); *Chem. Abstr.*, **65**, 674a (1966).

105   Akaboshi, S., and Kutsuma, T., *Yakugaku Zasshi*, **88**, 1011 (1968); *Chem. Abstr.*, **70**, 11524v (1969).

106   Tsukerman, S. V., Nikitchenko, V. M., Bugai, A. I., and Lavrushin, V. F., *Khim. Str., Svoistva Reaktivnost Org. Soedin.*, **53** (1969); *Chem. Abstr.*, **73**, 45276t (1970).

107   Samula, K., and Cichy, B., *Pol. J. Chem.*, **53**, 719 (1979); *Chem. Abstr.*, **91**, 157561k (1979).

108   Matsumura, K., Ito, M., and Lee, S. T., *J. Org. Chem.*, **25**, 854 (1960).

109   Akaboshi, S., and Kutsuma, T., *Yakugaku Zasshi*, **88**, 1016 (1968); *Chem. Abstr.*, **70**, 11525w (1969).

110   Funke, A., and Delavigne, R., *Bull. Soc. Chim. France*, 1974 (1959).

111   Krasnaya, Z. A., Stytsenko, T. S., Prokofév, E. P., and Kucherov, V. F., *Khim. Geterotsikl. Soedin.*, **5**, 668 (1973); *Chem. Abstr.*, **79**, 78541c (1973).

112   Mikhaleva, M. A., Gulevich, V. V., Naumenko, I. I., Mamaev, V. P., *Khim. Geterotsikl. Soedin.* (5), 678 (1979); *Chem. Abstr.*, **91**, 91592n (1979).

113   Mal'tseva, S. P., and Stepanov, B. I., *Izv. Vyssh. Ucheb. Zaved., Khim. Khim. Tekhnol.*, **16**, 1128 (1973); *Chem. Abstr.*, **79**, 105130b (1973).

114   Kucherenko, A. P., Potashnikova, S. G., Radkova, S. S., Baranov, S. N., Sheinkman, A. K., and Volbushko, N. V., *Khim. Geterotsikl. Soedin.*, (9), 1257 (1974); *Chem. Abstr.*, **82**, 72761q (1975).

115   Toma, S., and Perjessy, A., *Chem. Zvesti*, **23**, 343 (1969); *Chem. Abstr.*, **72**, 16942f (1970).

116   Marchenko, I. G., Turov, A. V., and Khilya, V. P., *Dopov. Akad. Nauk Ukr. RSR Ser. B* (1), 43 (1979); *Chem. Abstr.*, **90**, 152330d (1979).

117   Meyer, A., and Debard, R., *C. R. Acad. Sci., Paris, Ser. C*, **264**, 1775 (1967); *Chem. Abstr.*, **67**, 64503g (1967).

118   Reichel, L., and Doering, H. W., *Justus Liebigs Ann. Chem.*, **745**, 71 (1971).

119   Shah, N. M., *J. Univ. Bombay*, **11**, 109 (1942); *Chem. Abstr.*, **37**, 2000 (1943).

120   Yamada, S., Ono, F., Katagiri, T., and Tanaka, J., Bull. *Chem. Soc. Japan*, **48**, 2391 (1975).

121   Giri, S., and Singh, V. K., *Indian J. Chem.*, **14B**, 135 (1976).

122   Polyakov, V. K., Voronkin, V. M., and Tsukerman, S. V., *Ukr. Khim. Zh.*, **42**, 388 (1976); *Chem. Abstr.*, **85**, 21028k (1976).

123   Ansari, M. A., Mathur, K. B., and Ahluwalia, V. K., *Indian J. Chem.*, **16B**, 702 (1978).

124   Malik, S. B., Sharma, P., and Seshadri, T. R., *Indian J. Chem.*, **15B**, 519 (1977).

PART TWO
Reactions

Chapter Four

Reaction of Chalcones with Reducing Agents

CATALYTIC HYDROGENATION

Reduction of Olefinic Group[1-11]

Chalcone on catalytic hydrogenation with Raney nickel in ethanol yields benzylacetophenone.[1-4,10] The latter compound has been secured in high yield (96%) by employing dichloroethane in place of ethanol as a solvent.[5] High-pressure hydrogenation (50–120 atm) of chalcone in the presence of a nonpyrophoric Raney nickel catalyst (composition: copper, 58–60% and aluminum, 40–42%) is reported to yield dihydrochalcone (75%).[6]

Chalcones have been reduced by hydrogen and platinum to hydrochalcones,[7] which in turn can be reduced to hydrochalcols.

ortho-Hydroxychalcone[8] is reduced completely to the corresponding dihydrochalcone in ether in the presence of platinum black by a current of hydrogen. The preparation of several dihydrochalcones is described in the literature.[55-66]

With platinum oxide catalyst in ethanol (alkaline) the hydrogenation of chalcone stops with the absorption of one molecule of hydrogen and the exclusive formation of saturated ketone.[9] One study demonstrates the control of catalytic hydrogenation (catalyst: platinum oxide) of chalcone under the influence of added ferric chloride, different solvents, and temperature.[10] Thus the saturation of the ethylenic bond occurs very smoothly either in cold benzene, toluene, or boiling acetic acid.[10]

Pd_3B_2 is reported to be an efficient catalyst in the selective hydrogenation of the olefinic double bond in chalcone.[11] It has therefore been employed as a catalyst in the quantitative microhydrogenation of such types of compounds.[11]

Reduction of Carbonyl Group

p'-Methylchalcone is reduced by hydrogen to the corresponding unsaturated secondary alcohol[12] by using platinum black and a large excess of ferric chloride.

Selective reduction of the carbonyl group in chalcone has likewise been achieved by using an optimum amount of palladium catalyst (promoters: ferrous sulfate–zinc acetate) at ordinary atmospheric pressure and room temperature.[13] Hydrogenation of chalcone at atmospheric pressure, with colloidal palladium or with palladium precipitated on animal charcoal, is reported to reduce the carbonyl group smoothly.[14]

Reduction of Olefinic and Carbonyl Groups

1,3-Diphenylpropanol[5] has been secured by the catalytic hydrogenation of ethanolic solution of chalcone with Raney nickel. According to a report the same transformation has been accomplished in 45 minutes by incorporating traces of alkali in the reaction medium.[4]

Saturation of Olefinic Bond and Reduction of Carbonyl to Methylene Group

Catalytic reduction of chalcone by hydrogen in the presence of nickel (reduced at 250°, and partially deactivated at 200°) yields 1,3-diphenylpropane. A quantitative yield of this compound is obtained by the catalytic hydrogenation of chalcone with platinum catalyst[9] in ethanol in the presence of concentrated hydrochloric acid.

Reduction of Aromatic Rings and Enone Function

Chalcone undergoes hydrogenation in the presence of activated nickel catalyst at elevated temperature to give dicyclohexylpropane.[53] Formation of a similar perhydro compound is reported in the case of p'-methylchalcone, using platinum black as a catalyst.[12]

COMPLEX METAL HYDRIDE REDUCTION

Alkali Metal Borohydrides

Selective reduction of the double bond in chalcone occurs when it reacts in pyridine with sodium borohydride.[15]

Chalcone on treatment with potassium borohydride in water or methanol yields the corresponding carbinol.[16-18] This reaction has been extended for the preparation of several substituted allyl alcohols.[19,20] The selective reduction of the carbonyl group of chalcone has been achieved[17] by potassium borohydride and aluminum isopropoxide in isopropyl alcohol. The resulting compound, often referred to as chal-

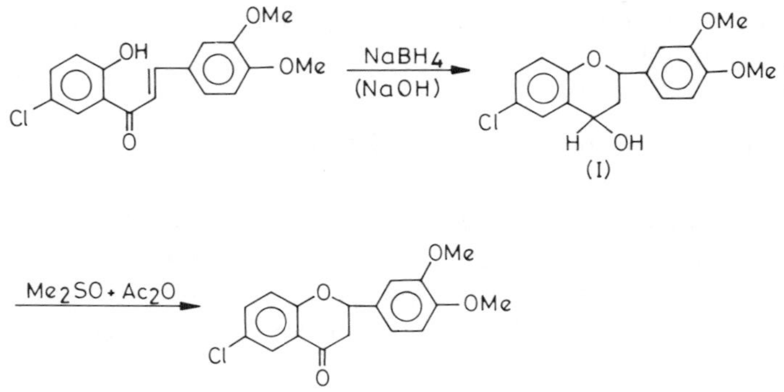

$\xrightarrow[\text{(NaOH)}]{\text{NaBH}_4}$

(I)

$\xrightarrow{\text{Me}_2\text{SO} + \text{Ac}_2\text{O}}$

col (1,3-diphenyl-2-en-1-ol), on treatment with alkali is reported[17] to undergo a rearrangement to yield the dihydrochalcone.

Sodium borohydride reduction of 2'-hydroxychalcone is reported to yield flav-3-ene besides the expected unsaturated alcohol.[21] 1,3-Diphenyl-1- (and 2-) propanols[22] are obtainable in 85% yield by carrying out reduction of chalcone with sodium borohydride–boron trifluoride, followed by treatment with alkaline hydrogen peroxide. Using chalcone as the starting material the synthesis of flavanol (I) and its corresponding flavanone has been achieved.[18]

The kinetics of reduction of chalcone with sodium borohydride has been reported.[23] A 1,4-addition mechanism has been suggested for this reduction.

LAH and Lithium Trimethoxyaluminohydride

The reduction of chalcone with LAH can be controlled to yield either the unsaturated[24] or saturated alcohol.[25] This reaction has been exploited in the synthesis of 4-cinnamylidene-2,5-cyclohexadien-1-ones.[26] Thus 4'-hydroxychalcones have been reduced by LAH to the corresponding unsaturated alcohols and the latter dehydrated to the aforesaid compounds. Chalcone is reported to give the unsaturated alcohol as the major product if LAH is replaced by the less aggressive reagent, Lithium trimethoxyaluminohydride.[27]

Lithium Aluminum Hydride–Aluminum Chloride (LAH–AlCl₃)

Treatment of chalcone with LAH–AlCl₃ yields the corresponding propenes[28] and a dimer, 1,3,5-triphenyl-4-benzyl-1-pentene[29] (44–65%).

The mechanism of the reaction has been postulated[28] as follows: With lithium aluminum hydride chalcone undergoes transformation to the saturated alcohol. The latter compound loses a molecule of water under the influence of the added Lewis acid, aluminum chloride, to yield the 1,3-diphenyl-1-propene.

In the hydrogenolysis of some chalcones (see below) two isomeric disubstituted propenes have been obtained.[28,30,31] In one instance the formation of an additional compound, 1-(o-hydroxyphenyl)-3-phenyl-1-propanone, has been reported.[30]

It is interesting to note that 2-hydroxy-4'-methoxychalcones on hydrogenolysis yield three products[28]: a saturated ketone (II), a secondary alcohol (III), and 4'-methoxyflavan (IV). Apparently the reduction follows a different course[28]:

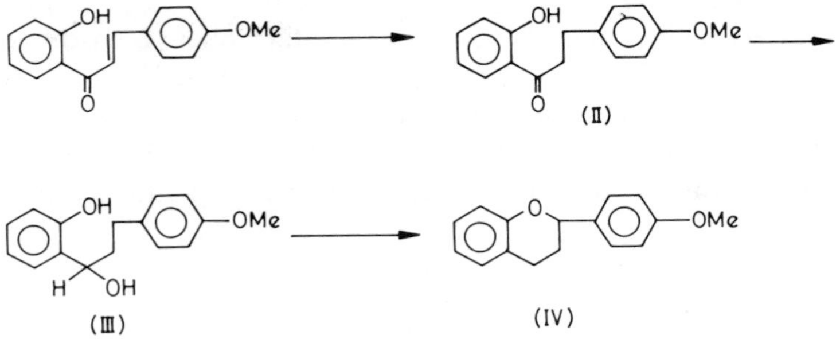

In the reduction of 4- and 4'-methoxychalcones by lithium aluminum hydride–aluminum chloride (1:2) two dimers are reported to be formed besides the isomeric 1,3-disubstituted propenes.[31] The following structures have been postulated for these dimers[31]:

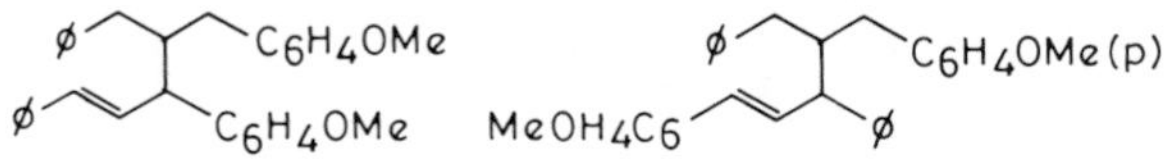

The behavior of α-(phenylsulfonyl)chalcone[32] toward mixed reagents (LAH–AlCl₃) is rather interesting. It results in the saturation of the ethylenic double bond, giving the corresponding hydrochalcone.[32]

Organotin Hydride Reduction

Diphenyltin hydride[33-35] brings about the reduction of chalcone to yield phenylstyrylcarbinol (75%). The distinctive feature of the reaction is that the two hydrogen atoms undergo uncatalyzed selective transfer from tin to the carbonyl group of chalcone to give the alcohol directly, and no hydrolysis step is required.

1,3-Diphenylpropan-1-one has been obtained from chalcone by its reaction with diphenyltin hydride under appropriate conditions. The saturation of the double bond is the net result of two steps: hydrostannation (1,4-addition) and hydrogenolysis.[36] With tributyltin,[36] however, the reaction does not proceed beyond the hydrostannation stage.

ELECTROLYTIC REDUCTION

Chalcone on electrolytic reduction is reported to yield three products: dibenzyldiacetophenone (35–40%), 1,3,4,6-tetraphenyl-3,4-dihydroxy-hexadiene (?, 25–30%), and the saturated ketone $\phi CH_2CH_2CO\phi$ (10–15%).[37] The yield of the latter could be raised to 70% by appropriate control of the reaction conditions. The following mechanism has been proposed for the electrolytic reduction of chalcone and its heterolytic analogue[38]:

$$R \cdot CH : CH \cdot COR' \xrightarrow[\text{Step (1)}]{e + \overset{\oplus}{H}}$$

$$\left[\begin{array}{c} R.CH:CH.\overset{\cdot}{C}(OH).R' \\ \updownarrow \\ R.\overset{\cdot}{C}H.CH:C(OH).R' \end{array} \right] \xrightarrow{\text{Step(2)}} R.\overset{\cdot}{C}H.CH_2COR'$$

$$\text{Step} \left| \begin{array}{c} e + \overset{\oplus}{H} \end{array} \right.$$
$$(3)$$
$$R.CH_2.CH_2COR'$$
$$(V)$$

$$\text{Step} \left| (2a) \right. \qquad \text{Step (3a)}$$

$$\text{Dimers}$$

(a) R = 2-quinolyl; R' = p-MeO—C_6H_4
(b) R = R' = phenyl
(c) R = 2-furyl; R' = phenyl

The ketone (V) can undergo further two one-electron reductions to yield the corresponding saturated secondary alcohol,[38] thus

$$R—CH_2—CH_2—COR' + e + H^+ \longrightarrow R—CH_2—CH_2—\overset{\cdot}{C}(OH)—R'$$
$$\downarrow e + H^+$$
$$R—CH_2—CH_2—CH(OH)—R'$$

MEERWEIN–PONNDORF–VERLEY REDUCTION

Allyl alcohols have been obtained (in 40–70% yield) by the Meerwein–Ponndorf–Verley reduction of chalcones.[39]

A modified method for the Meerwein–Ponndorf–Verley reduction of chalcone, using aluminum isopropoxide and isopropyl alcohol, has been reported.[40] The removal of acetone from the reaction mixture is unnecessary. The product, 1,3-diphenyl-2-propen-1-ol, is obtained in good yield (76%).

REDUCTION BY METALS

Lithium amalgam[41] brings about the reduction of chalcone, giving rise to a small amount of the corresponding alcohol (b.p. 168–71°). The reaction of chalcone with ammonium amalgam is not a clean one, and several products have been isolated.[54] The following are the products that are formed: 1,3,4,6-tetraphenylhexan-1,6-dione, 1,3,4,6-tetraphenyl-3,4-dihydroxy-1,5-hexadiene, 1,2,4,5-tetraphenylcyclohexan-1,2-diol, 1,3-diphenylpropyl alcohol, and 1-amino-1,3-diphenyl-3-propanol.

Benzylacetophenone is obtained from chalcone when the latter is treated with sodium in alcohol.[42]

1,3,4,6-Tetraphenylhexan-1,6-dione[42] is produced when chalcone is reduced as in the Clemmensen reaction or treated with zinc mercury in acetic acid.

2-Hydroxy-3,4-dimethoxy-3′,4′-methylenedioxychalcone on reduction with zinc in ethyl alcohol–acetic acid gives amorphous 4-hydroxy-7,8,4′-trimethoxyflavan.[38] Under the same reaction conditions butein yields the corresponding flavan or the pinacol[43]:

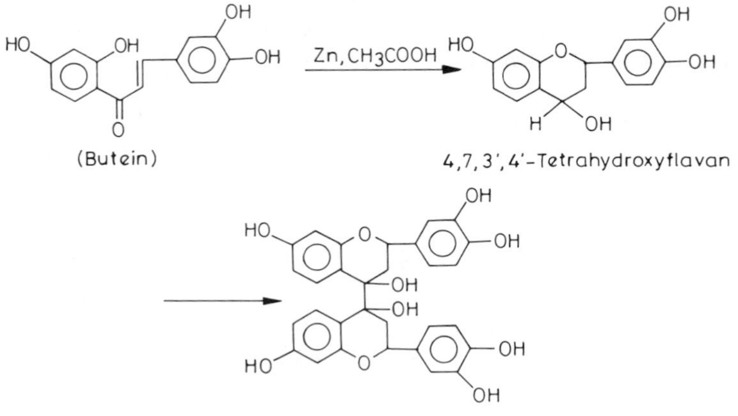

(Butein)   4,7,3′,4′-Tetrahydroxyflavan

Chalcone,[44] on the other hand, yields dibenzyldiacetophenone (m.p. 196°) and an isomer (m.p. 269°). Chalcone is reported to undergo reduction by vanadous and chromous salts with the formation of bimolecular products.[45] For example,

$$2\phi CH{=}CHCO\phi + 2V^{2+} + 2H^+ = (\phi CH{-}CH_2{-}CO{-}\phi)_2 + 2V^{3+}$$

WILLGERODT–KINDLER REACTION[46]

Dihydrochalcone (40%) has been obtained by heating chalcone with sulfur at 115° in the presence of quinoline. Besides the dihydrochalcone, 1,3-diphenylprop-1-ene is also formed when the above reaction is carried out at a higher temperature (145°). Reduction of olefinic bond in the Willgerodt–Kindler reaction has also been observed in the case of 4-chlorochalcone. However, according to a recent report,[67] cinnamic acid is produced when chalcone is heated in morpholine in the presence of sulfur.

REDUCTION BY "HANTZSCH ESTER"

The activated double bond of chalcone is selectively reduced[47] by "Hantzsch ester" (diethyl-1,4-dihydro-2,6-dimethylpyridine-3,5-dicarboxylate) to yield benzylacetophenone and illustrates a case of homogeneous hydrogen transfer reaction.

REDUCTION BY PYRIDINE–BORANE REAGENT

The selective reduction of the carbonyl group in chalcone has been accomplished with pyridine–borane reagent.[48]

$$\phi{-}CO{-}CH{=}CH{-}\phi \xrightarrow[\text{refluxing toluene, 4 hr}]{\text{Py–borane in}} \phi{-}CH(OH){-}CH{=}CH\phi$$
$$(84\%)$$

REDUCTION BY TERPENES[49]

Hydrogenation of the olefinic double bond in chalcone can be effected by the use of terpenes, especially phellandrene and limonene, as hydrogen donors.[49] Thus benzylacetophenone is produced when chalcone and limonene are refluxed in xylene. The same product results when *p*-chlorochalcone is subjected to the same treatment, using phellandrene as hydrogen donor. Apparently reductive dehalogenation occurs in the latter reaction.

MISCELLANEOUS REDUCTIONS

The reduction and alkylation in the α- and β-positions of chalcone in liquid ammonia have been reported.[50,51]

Chalcone has been converted into its corresponding saturated amine (53%) by reductive amination.[52] The reaction can be conducted by introducing hydrogen gas into an aqueous ethanolic ammonia solution containing chalcone and cyanocobalt complex.

REFERENCES

1 Bar, D., and Debruyne, Mme. E., *Ann. Pharm. Franc.*, **16**, 235 (1958); *Chem. Abstr.*, **53**, 4183h (1959).

2 Linke, H. A. B., and Eveleigh, D. E., *Z. Naturforsch.*, **30B**, 606 (1975).

3 Linke, H. A. B., and Eveleigh, D. E., *Z. Naturforsch.*, **30B**, 940 (1975).

4 Tucker, S. H., *J. Chem. Educ.*, **27**, 489 (1950).

5 Cornubert, R., and Eggert, H. G., *Bull. Soc. Chim. France*, 522 (1954).

6 Ponomarev, A. A., Finkel'shteĭn, A. V., and Kuz'mina, Z. M., *Zh. Obshh. Khim.*, **30**, 564 (1960).

7 Pfeiffer, P., Kalckbrenner, E., Kunze, W., and Levin, K., *J. Prakt. Chem.*, **119**, 109 (1928).

8 Bargellini, G., and Bini, L., *Gazz. Chim. Ital.*, **41**, II, 435.

9 Weidlich, H. A., and Delius, M. M., *Chem. Ber.*, **74B**, 1195 (1941).

10 Weygand, C., and Meusel, W., *Chem. Ber.*, **76B**, 498 (1943).

11 Burlaka, V. P., and Diakur, L. N., *Izv. Sib. Otd. Akad. Nauk SSSR Ser. Khim. Nauk* (2), 128 (1967); *Chem. Abstr.*, **68**, 53715x (1968).

12 Weygand, C., and Werner, A., *Chem. Ber.*, **71B**, 2469 (1938).

13 Csuros, Z., Zech, K., and Geczy, I., *Hung. Acta Chim.*, **1**, 1 (1946); *Chem. Abstr.*, **41**, 109i, 110a (1947).

14 Csuros, Z., *Muegyetemi Közlemények*, 110 (1947); *Chem. Abstr.*, **42**, 3727b (1948).

15 Iqbal, K., and Jackson, W. R., *J. Chem. Soc. C*, 616 (1968).

16 Huls, R., and Simon, Y., *Bull. Soc. Roy. Sci. Liege*, **25**, 89 (1956); *Chem. Abstr.*, **50**, 16664h (1956).

17   Davey, W., and Hearnes, J. A., *J. Chem. Soc.*, 4978 (1964).

18   Flammang, M., and Wermuth, C. G., *Bull. Soc. Chim. France* (2), Pt. (2), 674 (1973).

19   Lavrushin, V. F., Kutsenko, L. M., Grin, L. M., and Litvin, I. Y., *Ukr. Khim. Zh.*, **34**, 413 (1968); *Chem. Abstr.*, **69**, 76794g (1968).

20   Grin, L. M., Kutsenko, L. M., and Lavrushin, V. F., *Vestn. Khar'kov Univ.*, **46**, 81 (1970); *Chem. Abstr.*, **75**, 140426s (1971).

21   Cardillo, G., Cricchio, R., Merlini, L., and Nasini, G., *Gazz. Chim. Ital.*, **99**, 612 (1969).

22   Thankar, G. P., and Subbarao, B. C., *J. Sci. Ind. Res. (India)*, **21B**, 583 (1962).

23   Das, S. C., *Technology (Sindri, India)*, **11**, 91 (1974); *Chem. Abstr.*, **82**, 72350e (1975).

24   Hochstein, F. A., *J. Am. Chem. Soc.*, **71**, 305 (1949).

25   Brown, W. G., *Org. Reaction*, **6**, 482 (1954).

26   Kurosawa, K., Hashiba, A., and Takahashi, H., *Bull. Chem. Soc. Japan*, **51**, 3612 (1978).

27   Durand, J., and Huet, J., *Bull. Soc. Chim. France*, (7–8 Pt. 2), 428 (1978).

28   Bokadia, M. M., Brown, B. R., Cobern, D., Roberts, A., and Somerfield, G. A., *J. Chem. Soc.*, 1658 (1962).

29   Hase, T., *Acta Chem. Scand.*, **23**, 2409 (1969).

30   Hase, T., *Acta Chem. Scand.*, **22**, 2845 (1968).

31   Hase, T., *Acta Chem. Scand.*, **23**, 2403 (1969).

32   Balaiah, V., and Natarajan, C., *Indian J. Chem.*, **9**, 1025 (1971).

33   Kuivila, H. G., and Beumel, O. F., Jr., *J. Am. Chem. Soc.*, **80**, 3798 (1958).

34   Kuivila, H. G., and Beumel, O. F., Jr., *J. Am. Chem. Soc.*, **83**, 1246 (1961).

35   Kuivila, H. G., and Beumel, O. F., Jr., U.S. Patent, 2,997,485 (1958); *Chem. Abstr.*, **57**, 866f (1962).

36   Leusink, A. J., and Noltes, J. G., *Tetrahedron Letters*, 2221 (1966).

37   Shima, G., Mem. Coll. Sci. Kyoto Imp. Univ., **12A**, 327 (1929); *Chem. Abstr.*, **24**, 2118[5] (1930).

38   Ariyan, Z. S., Mooney, B., and Stonehill, H. I., *J. Chem. Soc.*, 2239 (1962).

39   Lavrushin, V. F., Kutsenko, L. M., and Grin, L. M., *Zh. Organ. Khim.*, **3**, 72 (1967); *Chem. Abstr.*, **66**, 94751h (1967).

40   Truett, W. L., and Moulton, W. N., *J. Am. Chem. Soc.*, **73**, 5913 (1951).

41   Kametani, T., and Nomura, Y., *J. Pharm. Soc. Japan*, **74**, 1037 (1954); *Chem. Abstr.*, **49**, 11594b (1955).

42   Frederick, J., Dippy, J., and Lewis, R. L., *Rec. Trav. Chim.*, **56**, 1000 (1937); *Chem. Abstr.*, **32**, 521[5,7] (1938).

43   Russell, A., *J. Chem. Soc.*, 218 (1934).

44   Finch, L. R., and White, D. E., *J. Chem. Soc.*, 3367 (1950).

45   Conant, J. B., and Cutter, H. B., *J. Am. Chem. Soc.*, **48**, 1016 (1926).

46   Compagnini, A., and Purrello, K., *Gazz. Chim. Ital.*, **95**, 686 (1965).

47 Braude, E. A., Hannah, J., and Linstead, R., *J. Chem. Soc.*, 3257 (1960).

48 Barner, R. P., Graham, J. H., and Taylor, M. D., *J. Org. Chem.*, **23**, 1561 (1958).

49 Kindler, K., and Luehrs, K., *Ann. Chem.*, **685**, 36 (1965); *Chem. Abstr.*, **63**, 8409h (1965).

50 Gautier, J. A., Miocque, M., and Duclos, J. P., *Bull. Soc. Chim. France* (12), 4348 (1969); *Chem. Abstr.*, **72**, 89494w (1970).

51 Gautier, J. A., Miocque, M., and Duclos, J. P., *Bull. Soc. Chim. France* (12), 4356 (1969); *Chem. Abstr.*, **72**, 100226h (1970).

52 Murakami, M., Suzuki, K., Fujishige, M., and Kang, J. W., *Nippon Kagaku Zasshi*, **85**, 235 (1964); *Chem. Abstr.*, **61**, 13408b (1964).

53 Frezouls, J., *Compt. Rend.*, **155**, 42.

54 Takagi, S., and Sakaguti, T., *J. Pharm. Soc. Japan*, **58**, 797 (1938); *Chem. Abstr.*, **33**, 2508[8] (1939).

55 Bargellini, G., and Finkelstein, M., *Gazz. Chim. Ital.*, **42**, II, 417; *Chem. Abstr.*, **7**, 1713[7] (1913).

56 Bargellini, G., and Martegiani, E., *Gazz. Chim. Ital.*, **42**, II, 427; *Chem. Abstr.*, **7**, 1713[9] (1913).

57 Bargellini, G., and Monti, L., *Gazz. Chim. Ital.*, **44**, II, 25 (1914); *Chem. Abstr.*, **9**, 447[9] (1915).

58 Bargellini, G., *Gazz. Chim. Ital.*, **44**, 421, *Atti. Accad. Lincei*, **23**, II, 135 (1914); *Chem. Abstr.*, **9**, 1042 (1915).

59 Massicot, J., *Compt. Rend.*, **240**, 94 (1955).

60 Bhat, A. R., Joshi, V., and Merchant, J. R., *Curr. Sci.*, **44**, 345 (1975).

61 Shinoda, J., Sato, S., and Kawagoe, M., *J. Pharm. Soc. Japan*, **49**, 797 (1929).

62 Klinke, P., and Gibian, H., *Chem. Ber.*, **94**, 26 (1961).

63 Maria Ada Jorio, *Ann. Chim. (Rome)*, **49**, 1929 (1959); *Chem. Abstr.*, **54**, 14235c (1960).

64 Bognár, R., Tokes, A. L., and Frenzel, H., *Acta Chim. (Budapest)*, **61**, 79 (1969); *Chem. Abstr.*, **71**, 113217k (1969).

65 Horowitz, R. M., and Gentili, B., U.S. Patent 3,429,873 (1969); *Chem. Abstr.*, **70**, 87315y (1969).

66 Laboratoires Coupin (S.A.), French Patent 1,221,869 (1960); *Chem. Abstr.*, **56**, P 10046i (1962).

67 Sathe, S. W., and Ghiya, B. J., *Curr. Sci.*, **46**, 446 (1977).

Chapter Five

Reaction of Chalcones with Oxidizing Agents

THALLIC SALTS

A simple and convenient method for the preparation of symmetrical[1] as well as unsymmetrical[1,2] benzils consists in the oxidation of appropriate chalcones with thallic nitrate (TN)[1]:

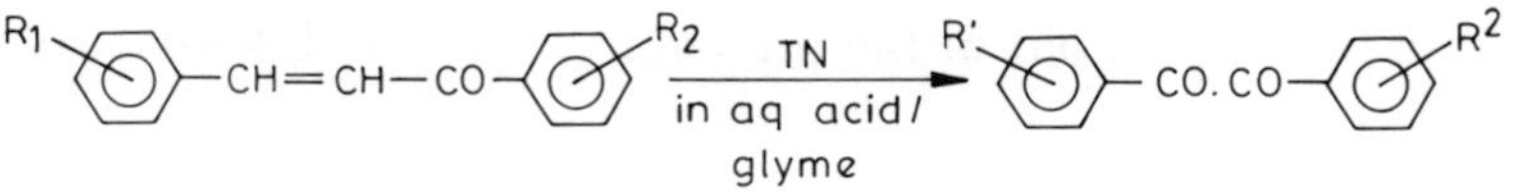

The yield of benzils is in the range of 45–70%. The reaction, however, fails when substituents prone to oxidation (viz., hydroxyl and amino groups) are present, or when both the aromatic rings carry electron-withdrawing functions.

The mechanism involved in the thallic nitrate (TN) oxidation of chalcone is shown below[1]:

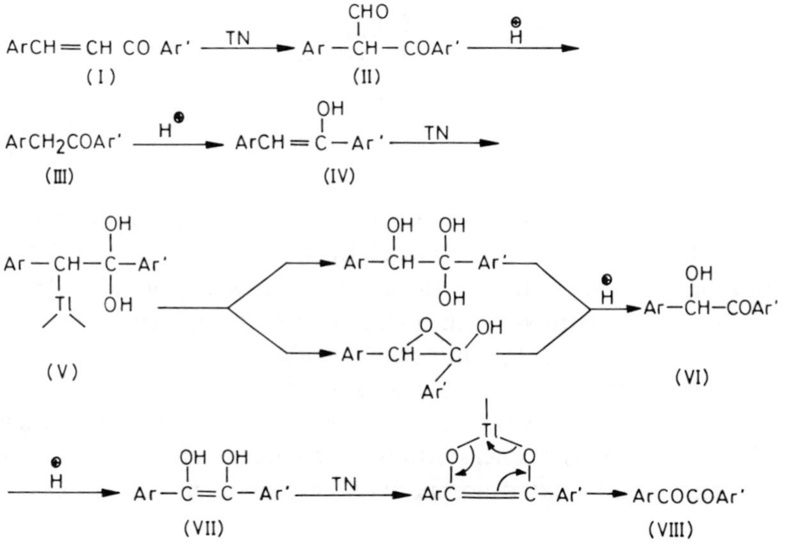

From the foregoing it is apparent that three distinct oxidations are involved in the overall reaction:

1   Chalcone (I) undergoes oxidative rearrangement to yield the ketoaldehyde (II), which, under acid catalysis, yields deoxybenzoin (III).[1]

2 Oxythallation of IV (enol tautomer of III), followed by a nucleophilic displacement of thallium, leads to the formation of benzoin (VI).

3 Finally, the oxidation of VI gives the benzil (VIII).

If the oxidative rearrangement of chalcone (IX) brought about by thallic salt is carried out in the presence of methanol, ketoacetal[3-6](X) results:

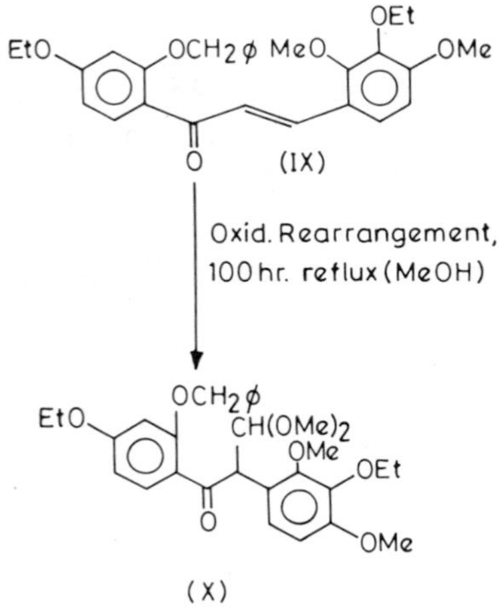

The reaction is assumed to involve the formation of the intermediate organothallium derivative,[8] followed by its rearrangement.[7,9,35,36] This reaction has been exploited for the synthesis of isoflavones.[7,9,35,36] For example, X, after treatment with methanolic hydrochloric acid, yielded smoothly 3',7-diethoxy-2',4'-dimethoxyisoflavone.[7]

2'-Hydroxychalcones[10] have also been shown to undergo oxidative rearrangement (see above) with thallic salts to yield finally the corresponding isoflavones.[10] The syntheses of several naturally occurring isoflavones[11-13] and related compounds are based on the aforesaid reaction.

The kinetics of thallic acetate oxidation of chalcone has been studied.[14]

LEAD TETRAACETATE

The β-hydroxychalcones on oxidation with lead tetraacetate[15] give the corresponding benzils together with aromatic acids. In the case of 4,4'-

dimethoxy-β-hydroxychalcone, however, the formation of three products, XI–XIII, is reported.[15]

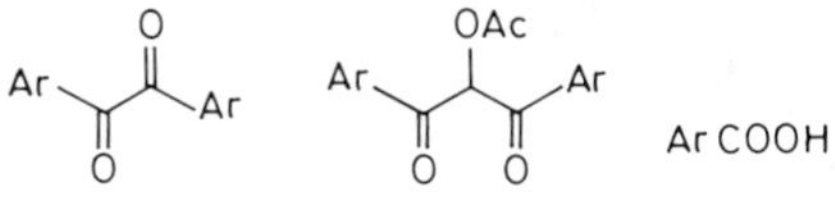

(XI) (XII) (XIII)

$Ar = p-CH_3O C_6H_4$

Lead tetraacetate oxidation of 2'-hydroxychalcones,[16] however, yield *cis*- and *trans*-aurones and substituted cinnamic and benzoic acids. Five products, **XIV–XVIII**, are obtained when 2'-benzyloxy-4,4'-dimethoxychalcone is oxidized by the above oxidant.

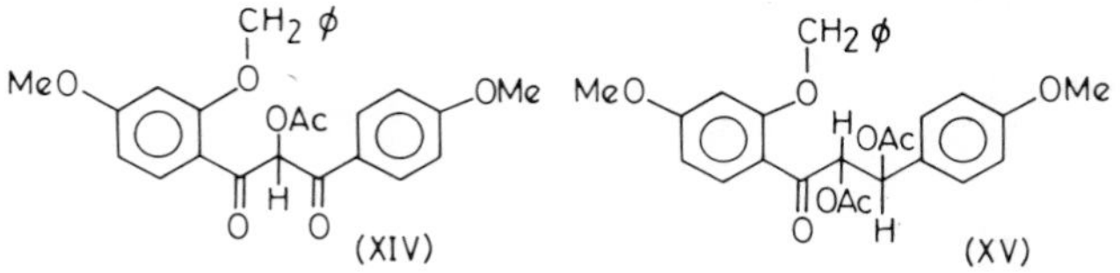

(XIV) (XV)

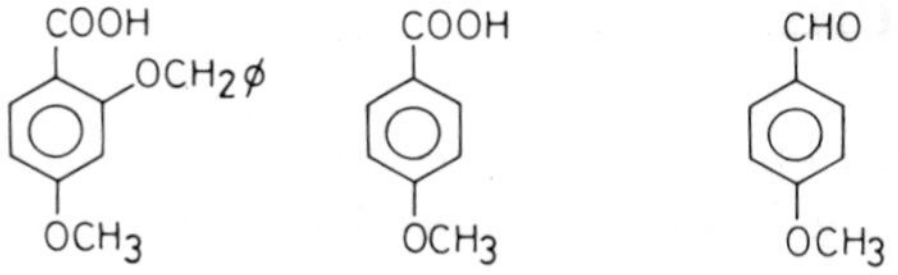

(XVI) (XVII) (XVIII)

SELENIUM DIOXIDE

Chalcones are oxidized smoothly with selenium dioxide to the flavones.[17-19] The following serves as an example for such a transformation[19]:

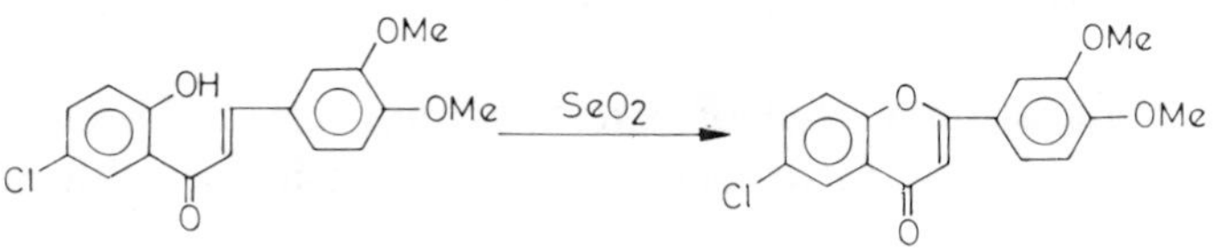

MANGANIC ACETATE

Fairly good yields of aurones[16,20] are obtainable from 2'-hydroxychalcones by oxidation with manganic acetate in acetic acid. The mechanism of the reaction is outlined as follows[20]:

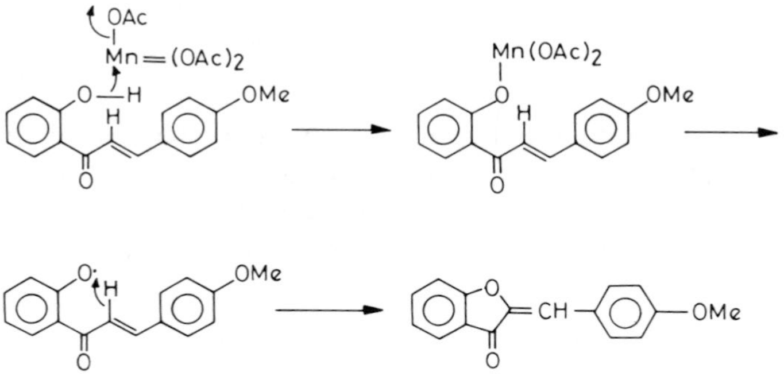

CHROMIC ACID

Epoxidation takes place when chalcone and its derivatives are treated with chromic acid.[21] The kinetics of this reaction has been studied.[21] The reaction is reported to involve the electrophilic attack by chromic acid at the olefinic center of the chalcone molecule, resulting in the formation of an epoxide.[21]

OSMIUM TETROXIDE[22]

Chalcone and osmium tetroxide react in ether solution to give a monoester, $C_{15}H_{12}O_5Os$, in 64% yield.

TRIFLUOROPEROXYACETIC ACID

Oxidative cyclization of chalcones to flavylium salts has been achieved by the use of trifluoroperoxyacetic acid[23] in methylene chloride. The reagent serves as an electrophilic hydroxylating system.[24] A plausible mechanism for this reaction is given here[23]:

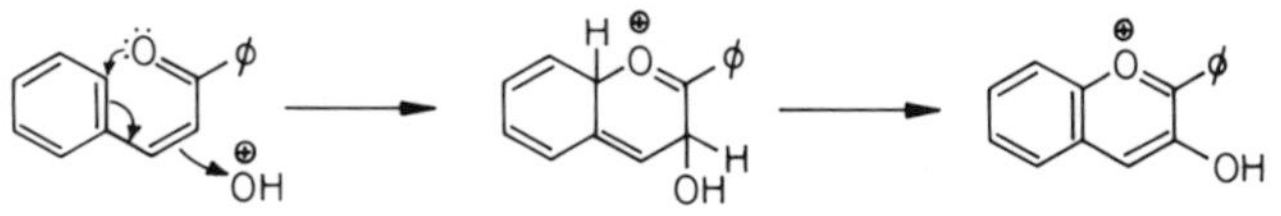

PERBENZOIC ACID

The reaction of perbenzoic acid with chalcone has been studied.[25] The oxidation of chalcone is considered to proceed through the intermediate formation of epoxy esters,[26] as follows:

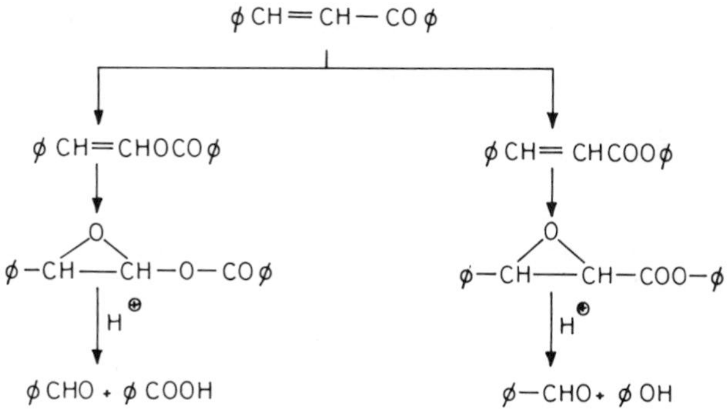

POTASSIUM PERSULFATE

Potassium persulfate, under appropriate experimental conditions, is reported to cause the nuclear oxidation of hydroxychalcones,[27] as is illustrated by 2'-hydroxy-4'-methoxychalcone:

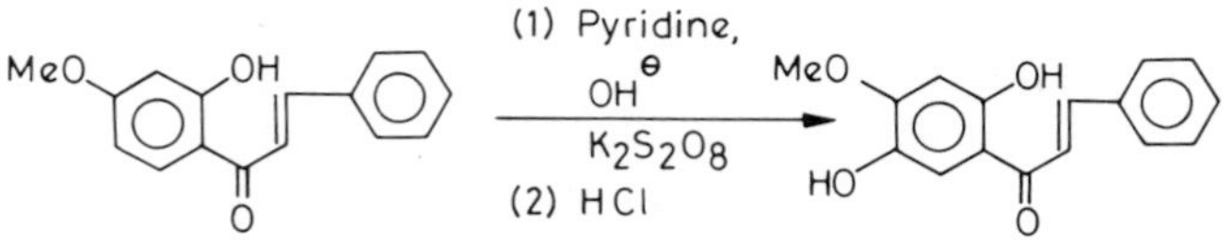

Some *trans*-1-aryl-2-aroyloxyethylenes (**XIX**) have been secured by the persulfate oxidation of chalcone and the naphthalene analogue of chalcone[28]:

$$\text{ArC}\!-\!\text{Ch}\!=\!\text{CH}\!-\!\text{Ar'} \xrightarrow{\text{K}_2\text{S}_2\text{O}_8-\text{H}_2\text{SO}_4-\text{HOAc}} \text{Ar}\!-\!\text{C}\!-\!\text{O}\!-\!\text{CH}\!=\!\text{CH}\!-\!\text{Ar'}$$

$$\overset{\|}{\text{O}} \qquad\qquad\qquad\qquad\qquad\qquad \overset{\|}{\text{O}}$$

(**XIX**)

Ar = Ph; 2-$C_{10}H_7$
Ar' = Ph; 4-FC_6H_4; 2-$C_{10}H_7$

tert-BUTYL HYDROPEROXIDE

Chalcone is reported to undergo epoxidation when treated with *tert*-butyl hydroperoxide[29] in the presence of Triton-B. With this reagent it is possible to carry out the reaction in a completely homogeneous non-polar medium. The mechanism involved in epoxidation is depicted below[29]:

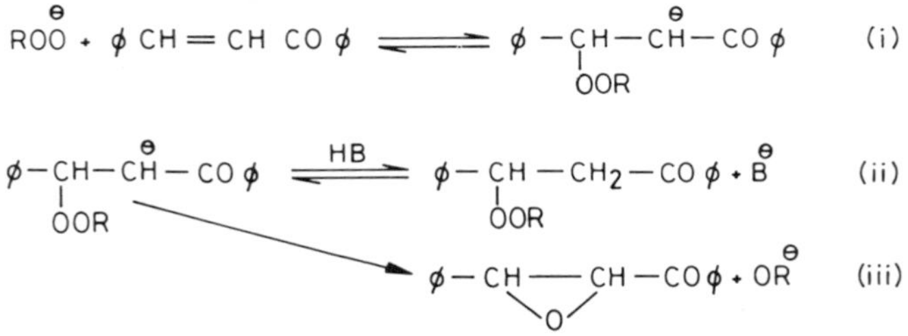

$$\overset{\ominus}{\text{ROO}} + \phi\,\text{CH}=\text{CH CO}\,\phi \rightleftharpoons \phi-\underset{\underset{\text{OOR}}{|}}{\text{CH}}-\overset{\ominus}{\text{CH}}-\text{CO}\,\phi \qquad (i)$$

$$\phi-\underset{\underset{\text{OOR}}{|}}{\text{CH}}-\overset{\ominus}{\text{CH}}-\text{CO}\,\phi \xrightarrow{\text{HB}} \phi-\underset{\underset{\text{OOR}}{|}}{\text{CH}}-\text{CH}_2-\text{CO}\,\phi + \overset{\ominus}{\text{B}} \qquad (ii)$$

$$\phi-\underset{\diagdown_{\!\!O}\diagup}{\text{CH}\!-\!\!-\!\text{CH}}-\text{CO}\,\phi + \overset{\ominus}{\text{OR}} \qquad (iii)$$

The equilibrium concentration of the carbanion is sufficiently large to enable step **iii** to compete with the protonation step **ii**. Since the elimination step is irreversible, the reaction is driven in the direction of epoxide formation.

Epoxidation of chalcones has also been achieved by using monoperphthalic acid,[30] as well as by hydrogen peroxide in an alkaline medium[31,32] (see under AFO reaction). The latter method has been extended to the preparation of heterocyclic epoxychalcones.[32]

AUTOXIDATION

Chalcone is autoxidized very slowly in the presence of potassium *tert*-butoxide–*tert*-butyl alcohol to 2 equiv of benzoic acid (75%).[33]

POTASSIUM FERRICYANIDE

2′,4-Dihydroxy-4′,6′-dimethoxychalcone (**XX**) is reported to undergo oxidative phenol coupling at the β-position to yield a diastereoiso-meric mixture of benzofuranones, **XXI** (25.5%) and aurone, **XXII** (23%).[34]

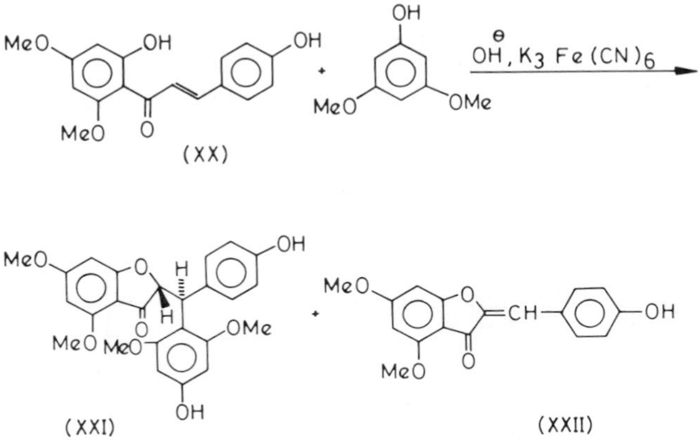

AMINE *N*-OXIDES

Flavones have been prepared from 2′-hydroxychalcones by their oxi-dation with amine *N*-oxides, viz., pyridine oxide and triethylamine ox-ide in the presence of a catalyst.[37]

REFERENCES

1   McKillop, A., Swann, B. P., and Taylor, E. C., *Tetrahedron Letters*, 5281 (1970).

2   McKillop, A., Swann, B. P., Ford, M. E., and Taylor, E. C., *J. Am. Chem. Soc.*, **95**, 3641 (1973).

3   Ollis, W. D., Ormand, K. L., and Sutherland, I. O., *J. Chem. Soc.*, *C*, 119 (1970).

4   Gottsegen, A., Antus, S., Farkas, L., Kardos-Balogh, Z., and Nagradi, M., *Fla-vonoids Bioflavonoids Proc. Hung. Bioflavonoid Symp., 5th, 1977*, pp. 181–185. Edited by T. Farkas, M. Gubor, and F. Kallay, Elsevier, Amsterdam, Nether-lands.

5   Taylor, E. C., Conley, R. A., Johnson, D. K., and McKillop, A., *J. Org. Chem.*, **42**, 4167 (1977).

6   Antur, S., Boross, F., Farkas, L., and Nogradi, M., *Flavonoid Bioflavonoids Proc.*

Hung, Bioflavonoid Symp., 5th, 1977, pp. 171–80. Edited by T. Farkas, M. Gubor, and F. Kalley, Elsevier, Amsterdam, Netherlands.

7   Ollis, W. D., Ormand, K. L., Redman, B. T., Roberts, R. J., and Sutherland, I. O., *J. Chem. Soc., C,* 125 (1970).

8   Ollis, W. D., Ormand, K. L., and Sutherland, I. O., *Chem. Commun.,* 1237 (1968).

9   Levai, A., and Balogh, L., *Pharmazie,* **30,** 747 (1975); *Chem. Abstr.,* **84,** 59070k (1973).

10   Farkas, L., Gottsegen, A., Nogradi, M., and Antus, S., *Chem. Commun.,* 825 (1972).

11   Farkas, L., Gottsegen, A., Nogradi, M., and Antus, S., *J. Chem. Soc., Perkin Trans.,* **1,** 305 (1974).

12   Farkas, L., Antus, S., and Nogradi, M., *Acta Chim. Acad. Sci. Hung.,* **82,** 225 (1974); *Chem. Abstr.,* **81,** 120319v (1974).

13   Antus, S., Farkas, L., Kardos-Balogh, Z., and Nogradi, M., *Chem. Ber.,* **108,** 3883 (1973).

14   Khandual, N. C., Satpathy, K. K., and Nayak, P. L., *Proc. Indian Acad. Sci.,* **76A,** 129 (1972).

15   Kurosawa, K., and Moriyama, A., *Bull. Chem. Soc. Japan,* **47,** 2717 (1974).

16   Kurosawa, K., and Higuchi, J., *Bull. Chem. Soc. Japan,* **45,** 1132 (1972).

17   Mahal, H. S., Rai, H. S., and Venkataraman, K., *J. Chem. Soc.,* 866 (1935).

18   Mahal, H. S., and Venkataraman, K., *J. Chem. Soc.,* 569 (1936).

19   Chakravarti, D., and Dutta, J., *J. Indian Chem. Soc.,* **16,** 639 (1939).

20   Kurosawa, K., *Bull. Chem. Soc. Japan,* **42,** 1456 (1969).

21   Khandual, N. C., Satpathy, K. K., and Nayak, P. L., *J. Chem. Soc. Perkin Trans.,* **2,** 328 (1974).

22   Criegee, R., Marchand, B., and Wannowius, H., *Annalen,* **550,** 99 (1942).

23   Brown, B. R., Davidson, A. J., and Normon, R. O. C., *Chem. Ind. (London),* 1237 (1962).

24   Chambers, R. D., Goggin, P., and Musgrave, W. K. R., *J. Chem. Soc.,* 1804 (1959).

25   Young, W. G., Cristol, S. J., Andrew, L. J., and Lindenbaum, S. L., *J. Am. Chem. Soc.,* **66,** 855 (1944).

26   Yokoyama, T., and Nohara, F., *Bull. Chem. Soc. Japan,* **38,** 1498 (1965).

27   Rajagopalan, S., and Seshadri, T. R., *Proc. Indian Acad. Sci.,* **27A,** 85 (1948).

28   Dhar, D. N., and Munjal, R. C., *Synthesis,* **9,** 542 (1973).

29   Yang, N. C., and Finnegan, R. A., *J. Am. Chem. Soc.,* **80,** 5845 (1958).

30   Verma, B. L., and Bokadia, M. M., *J. Indian Chem. Soc.,* **42,** 399 (1965).

31   Belyaev, V. F., and Yatsevich, N. M., *Khim. Geterotsikl. Soedin.,* **4,** 582 (1968); *Chem. Abstr.,* **70,** 37565j (1969).

32   Korotkov, S. O., Orlov, V. D., and Lavrushin, V. F., *Vestn. Kharkov Univ.,* **84,** 72 (1972); *Chem. Abstr.,* **78,** 147695z (1973).

33 Doering, W. V. E., and Haines, R. M., *J. Am. Chem. Soc.*, **76**, 482 (1954).

34 Volsteedt, F. du R., Ferreira, D., and Roux, D. G., *Chem. Commun.*, 217 (1975).

35 Farkas, L., and Wolfner, A., *Acta Chim. Acad. Sci. Hung.*, **88**(2), 173 (1976).

36 Jain, A. C., Khazanchi, R., and Gupta, R. C., *Bioorg. Chem.*, **7**, 493 (1978).

37 Haginiwa, J., Higuchi, Y., Kawashima, T., and Shinokawa, H., *Yakugoku Zasshi*, **96**, 195 (1975); *Chem. Abstr.*, **85**, 123724b (1976).

Chapter Six

Reaction of Chalcones with Ketones

ISOPROPYL AND ISOBUTYL KETONES

Several diketones, such as *erythro–threo* mixtures, have been prepared by the reaction of chalcones with enolates derived from isopropyl and isobutyl ketones.[1]

ACETOPHENONES

Chalcone is reported to undergo 1,2- or 1,4-addition with the ketone enolates, depending on the conditions of the experiment.[2] Chalcone and MeO–CH$_2$COPh cyclized with ammonium acetate–acetic acid to yield 2,4,6-triphenylpyridine[3]. The chalcone–LAH complex in ether solution is reported to react with acetophenone to yield 2,3,5-triphenylpentan-2,5-diol.[4]

BENZOPHENONE

The benzophenone-sensitized irradiation of chalcone furnishes 2-benzoyl-3,4,6-triphenyl-1,2,3-dihydropyran[5] (**I**) in accordance with the following mechanism:

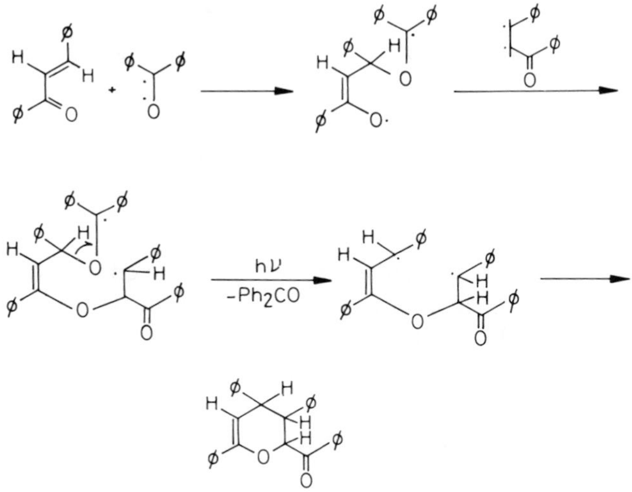

(I)

Disodiobenzophenone salt is reported to react with chalcone to form, after acidification, the keto alcohol, β,γ,γ-triphenyl-γ-hydroxy-butyrophenone in 47% yield.[6]

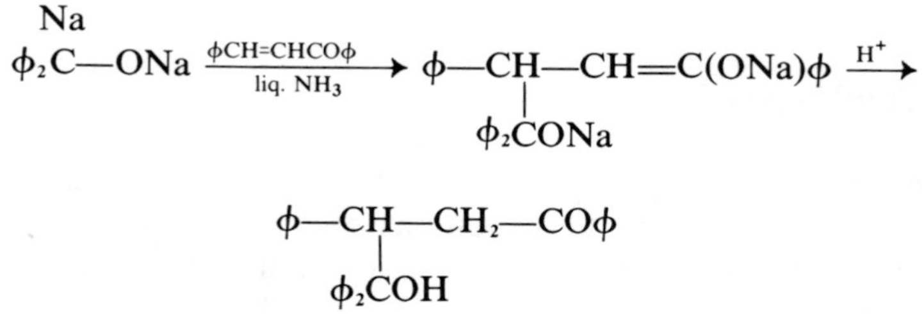

$$\overset{\text{Na}}{\phi_2 C}\!\!-\!\!ONa \xrightarrow[\text{liq. NH}_3]{\phi CH=CHCO\phi} \phi\!-\!\underset{\underset{\phi_2CONa}{|}}{CH}\!-\!CH\!=\!C(ONa)\phi \xrightarrow{\text{H}^+}$$

$$\phi\!-\!\underset{\underset{\phi_2COH}{|}}{CH}\!-\!CH_2\!-\!CO\phi$$

PHENYLBENZYL KETONE[7]

A quantitative yield of the Michael adduct is obtained when 4'-methylchalcone is treated with phenylbenzyl ketone under alkaline conditions. However, a pyrylium salt[8] is produced if triphenylmethyl perchlorate is used in the aforesaid reaction. Thus

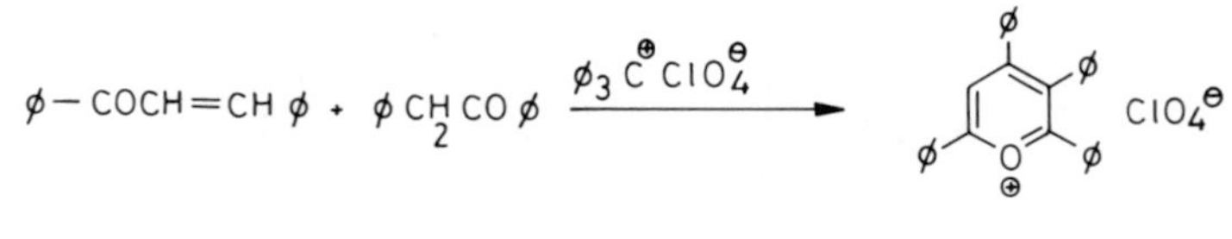

CYCLOPENTANONE[9-12]

Under the influence of diethylamine or piperidine chalcone reacts with cyclopentanone, yielding a semicyclic-1,5-diketone, α-benzoyl-β-phenylcyclopentanonylethane[9] (II):

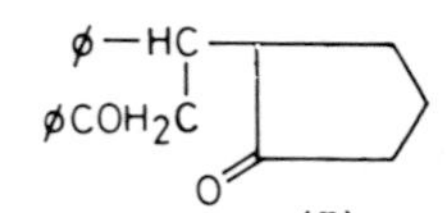

(II)

3-METHYLCYCLOHEXANONE[11,12]

The 1,5-diketone, α-benzoly-β-phenyl-(3-methylcyclohexanonyl)ethane, in two stereoisomeric forms, is obtained by the reaction of 3-methylcyclohexanone with chalcone under appropriate conditions. Michael reaction of chalcone with 2-methylcyclohexanone is also reported.[10]

TETRAPHENYLCYCLOPENTADIENONE

2,3,4,5,6-Pentaphenylbenzophenone[13] results from the reaction of chalcone and tetraphenylcyclopentadienone at an elevated temperature ($\sim 300°$).

MENTHONE

A bicyclic ketoalcohol,[14] III or IV, results from the interaction of chalcone and menthone in the presence of sodium ethoxide:

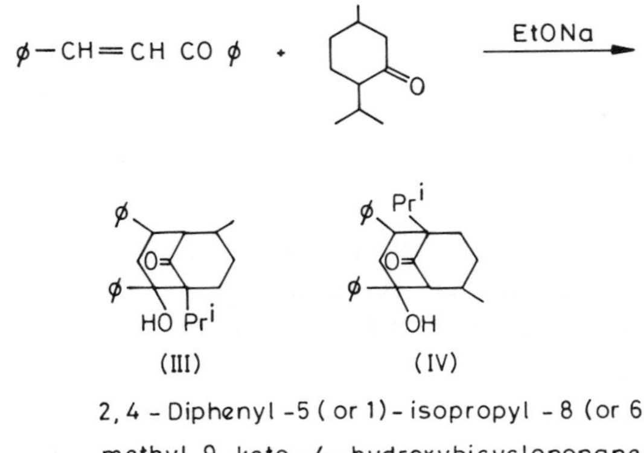

2,4 – Diphenyl –5 (or 1)– isopropyl – 8 (or 6)
methyl 9–keto–4–hydroxybicyclononane

FENCHONE AND CAMPHOR

Dibenzaltriacetophenone[15] is formed when chalcone reacts with alcoholic sodium hydroxide in the presence of fenchone or camphor.

FLAVANONES[16,17]

Chalcone and flavanone undergo addition reaction in the presence of a base—sodium amide, sodium, or caustic alkali—to yield 2-phenyl-2-phenacylbenzyl-2,3-dihydro-1,4-benzopyrone.

The presence of a 3',4'-methylenedioxy substituent in the chalcone component retards the aforesaid reaction, while a nitro group inhibits the same.[17] A mechanism for the aforesaid reaction has been suggested.[18]

2-SUBSTITUTED 5,6-BENZOCHROMANONE

Michael adducts are formed when chalcone is allowed to react with 2-substituted 5,6-benzochromanone.[19] The following serves as an illustrative example:

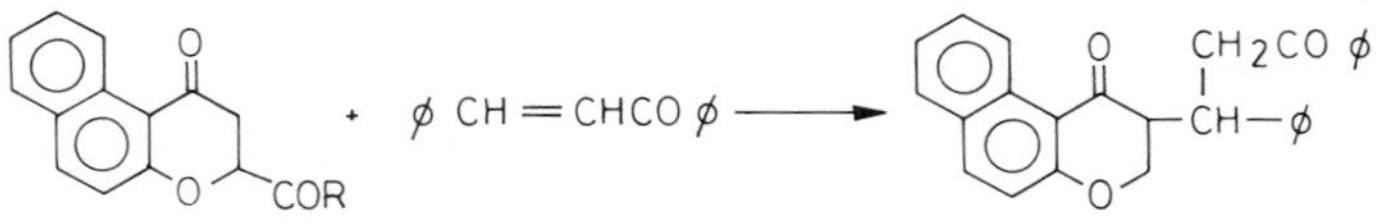

R = OH

ACETYLACETONE

Chalcone reacts with acetylacetone[20] (at 120°) to yield a cyclohexenone derivative (V). When the above reaction is conducted at room temperature an addition product (VI) is obtained[20]:

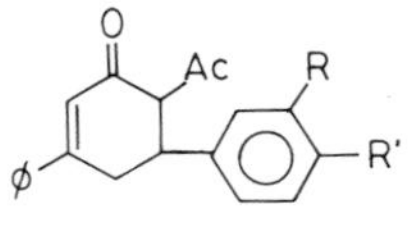

$\phi - COCH_2 CH(C_6H_3RR'-3,4)CH(COMe)_2$

VI (60%)

V (40 – 43%)

[R = R'= H ; R = H ; R'= OMe ; RR'= OCH_2O]

DIHYDRORESORCINOL[21]

2-[1-Phenyl-2-benzoylethyl]cyclohexan-2,6-dione **(VII)** has been secured in 58% yield by the reaction of dihydroresorcinol and chalcone.

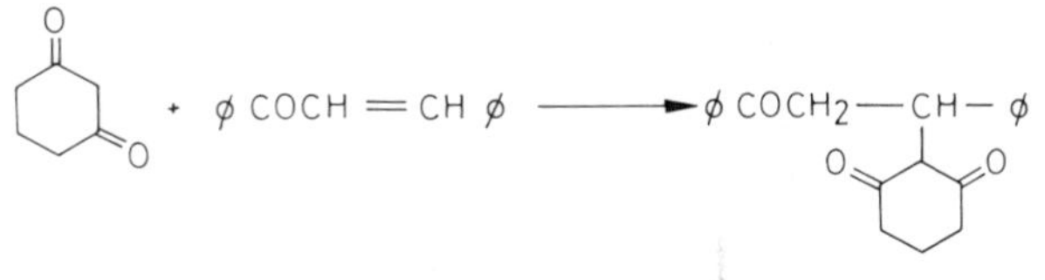

(VII)

5,5-DIMETHYLCYCLOHEXANDIONE

Condensation of **VIII** with chalcone or *pp'*-dimethoxychalcone in the presence of piperidine or sodium ethoxide gave **IX**[22] in a good yield. Higher yields were obtained with sodium ethoxide catalyst.

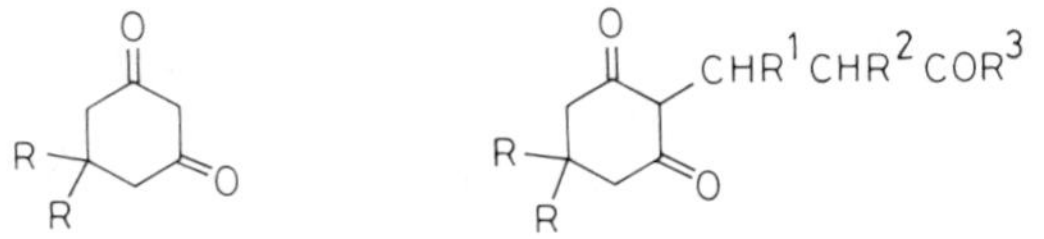

(VIII)

(R = CH₃)

(IX)

(R¹, R³ = C₆H₅ ; P−MeO C₆H₄ ; R² = H)

2-NITRO-1,3-INDANDIONE

A good yield (74%) of 1,3-diphenyl-3-(2-nitro-1,3-indandion-2-yl)-1-propanone has been obtained by the interaction of chalcone with 2-nitro-1,3-indandione in hexane.[23]

BENZYLIDENE BIS(ACETOPHENONE)

The synthesis of pyrylium salt has been achieved by the interaction of benzylidene bis(acetophenone) with chalcone in acetic acid medium, using boron trifluoride as the catalyst.[24]

1,3,5-TRIPHENYLPENTAN-1,5-DIONE

The interaction of 1,3,5-triphenylpentan-1,5-dione and chalcone in the presence of perchloric acid results in the formation of pyrylium salt.[25] In this reaction chalcone is reported to act as a hydride ion acceptor.[25]

REFERENCES

1 Gorrichon-Guigon, L., Maroni-Barnaud, Y., and Maroni, P., *Bull. Soc. Chim. France* (11), 4187 (1972).

2 Bertrand, J., Gorrichon, L., and Maroni, P., *Tetrahedron Letters*, 4207 (1977).

3 Teuber, H. J., Bader, H. J., and Schuetz, G., *Justus Liebigs Ann. Chem.*, 9(8), 1335 (1977).

4 Dornow, A., and Bartsch, W., *Chem. Ber.*, **87**, 633 (1954).

5 Abdel-Megeid, E. M. F., Bose, A. K., Elkaschef, A. F., Elsayed, A. S., Mokhtar, K. E., and Sharma, S. D., *Indian J. Chem.*, **13**, 482 (1975).

6 Hamrick, P. J., Jr., and Hauser, C. R., *J. Am. Chem. Soc.*, **81**, 493 (1959).

7 Sammour, A., Selim, M. I. B., Essawy, A., and Elkasaby, M., *Egypt. J. Chem.*, **16**, 197 (1973); *Chem. Abstr.*, **81**, 135630y (1974).

8 Carretto, J., Sib, S., and Simalty, M., *Bull. Soc. Chim. France* (6), 2312 (1972).

9 Striegler, C., *J. Prakt. Chem.*, **86**, 241.

10 Sammour, A., Raouf, A., Elkasaby, M., and Ibrahim, M. A., *Acta Chem. (Budapest)*, **78**, 399 (1973); *Chem. Abstr.*, **79**, 146100k (1973).

11 Georgi, R., and Volland, H., *J. Prakt. Chem.*, **86,** 232.

12 Rosenberg, A., *J. Prakt. Chem.*, **86**, 250.

13 Wolinski, J., *Roczniki Chem.*, **26**, 168 (1952); *Chem. Abstr.*, **49**, 6873e (1955).

14 Stobbe, H., and Rosenberg, A., *J. Prakt. Chem.*, **86**, 226.

15 Georgi, R., and Schwyzer, A., *J. Prakt. Chem.*, **86**, 273.

16 Kaplash, B. N., Shah, R. C., and Wheeler, T. S., *Curr. Sci.*, **8**, 512 (1939).

17 Kaplash, B. N., Shah, R. C., and Wheeler, T. S., *J. Indian Chem. Soc.*, **19**, 117 (1942).

18 Dhar, D. N., *Agra Univ. J. Res. (Sci.) Part I*, **10**, 75 (1961); *Chem. Abstr.*, **58**, 432g (1963).

19 Sammour, A., Selim, M. I. B., Elkasaby, M., and Saied, F., *J. Prakt. Chem.*, **314**, 941 (1972).

20 Sammour, A., Selim, M. I. B., and Hataba, A. M., *Egypt. J. Chem.*, **15**, 531 (1972); *Chem. Abstr.*, **80**, 82181y (1974).

21 Mikhailov, B. M., *J. Gen. Chem. (USSR)*, **7**, 2950 (1937).

22 Lelyukh, L. I., Korshunova, K. M., and Kharchenko, V. G., *Izv. Vyssh. Ucheb. Zaved., Khim. Khim. Tekhnol.*, **18**, 1536 (1975); *Chem. Abstr.*, **84**, 30597b (1976).

23 Zalukaevs, L. P., and Klykova, L. V., *Zh. Obshch. Khim.*, **34**, 3821 (1964).

24 Csuros, Z., Sallay, P., and Deak, G., *Acta Chim.* (*Budapest*), **79**, 349 (1973); *Chem. Abstr.*, **80**, 59143d (1974).

25 Van Allan, J. A., and Reynolds, G. A., *J. Org. Chem.*, **33**, 1102 (1968).

Chapter Seven

Reaction of Chalcones with Esters

ETHYL ACETATE

Chalcone undergoes a Michael type of reaction with ethyl acetate[1] in the presence of triphenylmethylsodium to yield ethyl-α-acetyl-β-phenyl-γ-benzoylbutyrate (66%).

ETHYL ACETOACETATE

Chalcone reacts with ethyl acetoacetate to yield a variety of different compounds, viz., Michael adduct,[2] pyrylium salt,[3] or cyclohexanone[4] derivative, depending on the experimental conditions employed. Thus in the presence of boron trifluoride etherate the above reaction yields the pyrylium salt according to the reaction[3]:

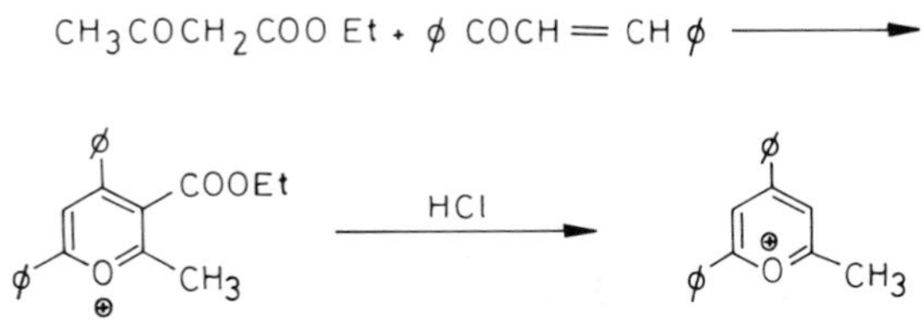
CH3COCH2COO Et + ϕ COCH=CH ϕ
COOEt
CH3
HCl
CH3

Here the chalcone, besides entering the reaction, also acts as a hydride ion acceptor and is converted to dihydrochalcone. Chalcone[4] and its derivatives[5–7] undergo condensation with ethyl acetoacetate to yield the corresponding cyclohexanone derivatives. For example,[4,8]

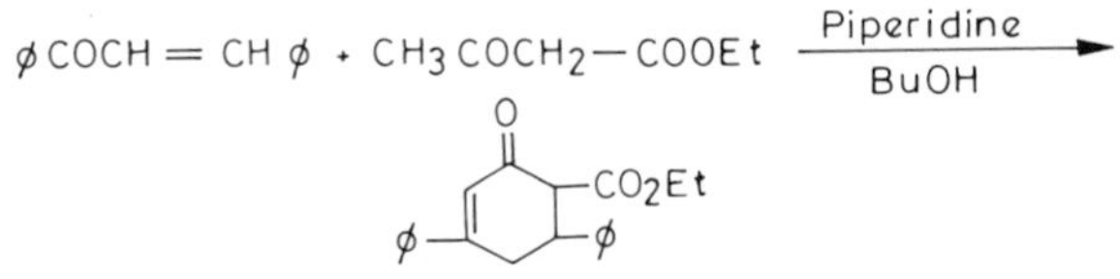
ϕCOCH=CH ϕ + CH3COCH2—COOEt
Piperidine
BuOH
CO2Et

The furan analogue of chalcone[9] also reacts with ethyl acetoacetate in the above manner.

ETHYL PHENYLACETATE

Chalcone-carrying substituents in ring B react with ethyl phenylacetate[10] to yield butyrates of the type $R'{-}C_6H_4CH(CH_2COPh)CHPh{-}CO_2Et$. Under Michael conditions α-phenylchalcones[11] react with ethyl phenylacetate to yield the adducts.

ETHYL BENZOYLACETATE

Chalcones react with ethyl benzoylacetate[12] to yield the Michael adduct, which on treatment with acid undergoes decarbethoxylation to furnish the diketone:

$$4\text{—MeO—C}_6\text{H}_4\text{—}\overset{\overset{\displaystyle O}{\|}}{\text{C}}\text{—CH=CH}\phi + \text{C}_6\text{H}_5\text{COCH}_2\text{—}\overset{\overset{\displaystyle O}{\|}}{\text{C}}\text{—OEt} \xrightarrow{\text{alc. NaOH}}$$

$$4\text{—MeO—C}_6\text{H}_4\text{—}\overset{\overset{\displaystyle O}{\|}}{\text{C}}\text{—CH}_2\text{—CH}\phi\text{—CH}\underset{\text{CO}\phi}{\overset{\text{COOEt}}{<}} \xrightarrow{\text{HOAc–HCl}}$$

$$4\text{—MeOC}_6\text{H}_4\text{COCH}_2\text{CH}\phi\text{CH}_2\text{CO}\phi$$

ETHYL CYANOACETATE[4,13-15]

Chalcones and ethyl cyanoacetate have been found to react in the following way:

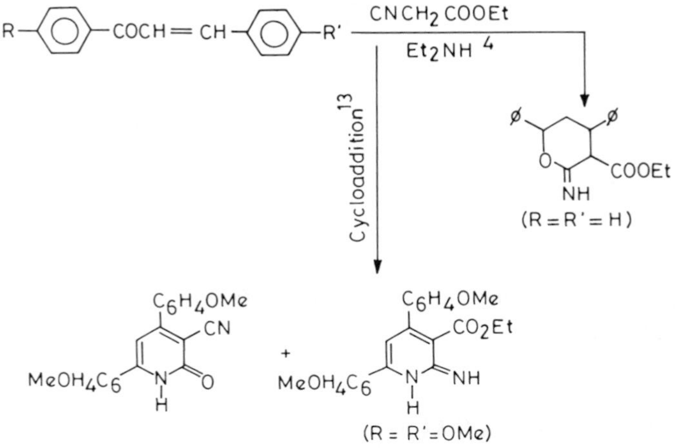

ETHYL α-CYANOBUTYRATE

Chalcone undergoes Michael addition when it reacts with ethyl α-cyanobutyrate. The kinetics of this reaction has been investigated.[16]

ETHYL THIOGLYCOLATE

Chalcone and nitrochalcones are reported to react with ethyl thiogly-colate in the presence of piperidine to yield an addition compound (I) and a cyclized product (II).[17]

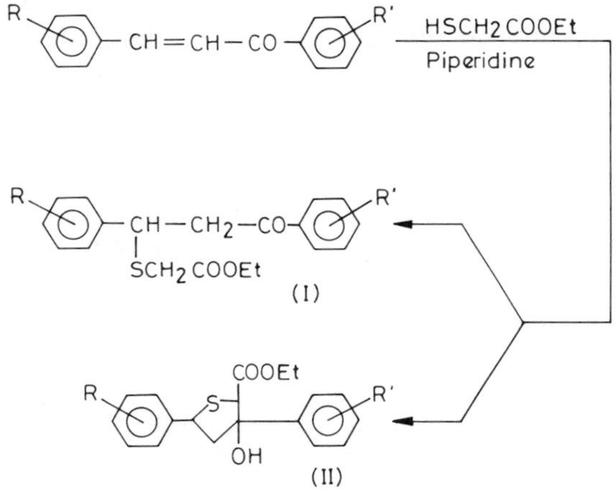

i) R = R' = H

ii) R = H ; R' = 2 - (3- and 4 -) NO_2

MALONIC ESTERS

Malonic ester[2,18] undergoes Michael addition with chalcone in the presence of a basic catalyst, piperidine. The mechanism of this reaction is reported.[18] Polyfluorochalcones[19] also undergo the above-mentioned reaction, but with difficulty.

Chalcone has been found to react with malonic ester[20] in a basic medium (or in the presence of a complex derived from ketal and po-

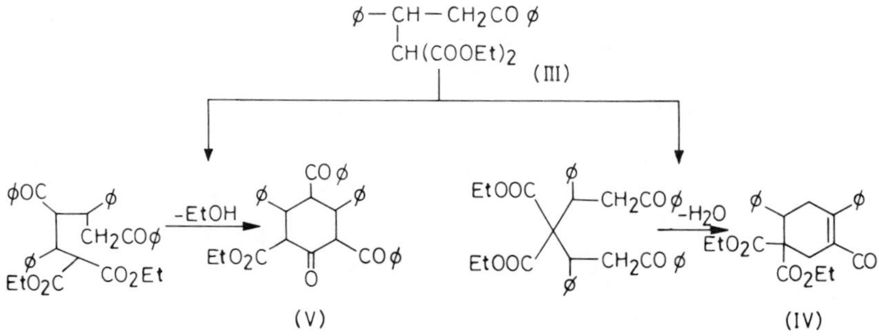

tassium hydroxide)[21,22] to yield α-carbethoxy-β-phenyl-γ-benzoylbu-
tyric ester[21,23] (III). Depending, however, on the conditions of the
experiment, other products, IV and V, can also be obtained.[23,24]

Methylmalonic ester[23,25] likewise adds to chalcone to yield α-meth-
yl-α-carbethoxy-γ-phenyl-γ-benzoylbutyric ester[23] (VI, 80%) with a
larger concentration of ethoxide (1 M); however, VI undergoes com-
plete retrogression[23] to yield sodium enol methylmalonic ester (90%),
chalcone, and traces of benzoylacetic ester. Butyric ester, VI, obtained
as above but in the presence of piperidine, undergoes further reaction
as illustrated[26]:

$$
\begin{array}{ccc}
\text{Ph—CH—CH(COOEt)COPh} & & \text{PhCH=\!C(CH}_3)\text{COOEt} \\
| & \longrightarrow & + \\
\text{CH}_3\text{—CH—COOEt} & & \text{PhCOCH}_2\text{COOEt}
\end{array}
$$

Ethyl alkyl malonate[27] and chalcone undergo Michael condensation to give the product (alkyl=CH$_3$; 80%). In the other case
(alkyl=ethyl) the expected Michael condensation product is not iso-
lable. This has been attributed to steric hindrance as well as to the in-
stability of the Michael addition product toward sodium ethoxide
present in the reaction medium.[27]

METHYL SUCCINATE

Methyl succinate reacts with chalcone, yielding different products[7,12]
depending on the reaction conditions. Thus

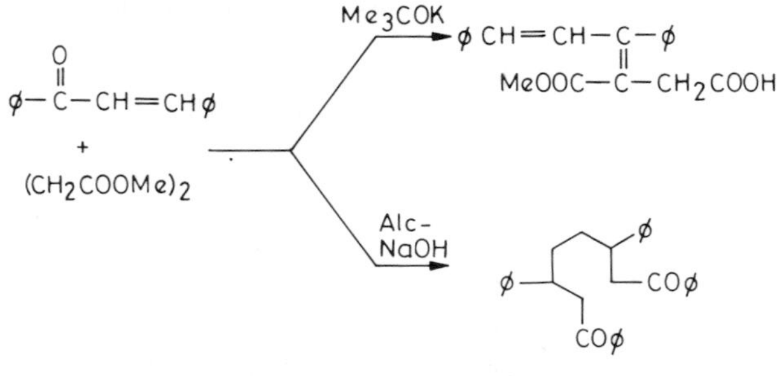

1,3,6,8-Tetraphenyl-1,8-octandione

DIMETHYLACETYLENE DICARBOXYLATE[28]

2-Hydroxychalcone and 2,2'-dihydroxychalcone react with dimethyla-cetylene dicarboxylate (DMAD) to give a mixture of phenoxymaleates (VII), phenoxyfumarates (VIII), and some cyclic products, viz., chromenes (IX) and/or flavanones (X). The reaction of 3,5-dibromo-2,2'-dihydroxychalcone with DMAD is typical and is illustrated as follows:

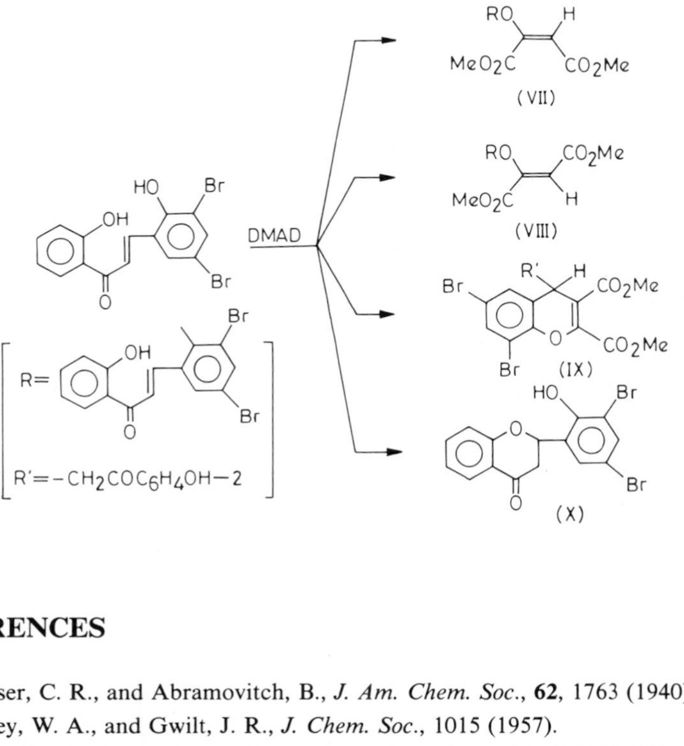

REFERENCES

1 Hauser, C. R., and Abramovitch, B., *J. Am. Chem. Soc.*, **62**, 1763 (1940).

2 Davey, W. A., and Gwilt, J. R., *J. Chem. Soc.*, 1015 (1957).

3 Van Allan, J. A., and Reynolds, G. A., *J. Org. Chem.*, **33**, 1102 (1968).

4 Sammour, A-M. A., El-Zimaity, M. T., and Abdel-Maksoud, A., *J. Chem. U.A.R.*, **12**, 481 (1969); *Chem. Abstr.*, **74**, 42098w (1971).

5 Merchant, J. R., and Choughuley, A. S. U., *Curr. Sci. (India)*, **29**, 93 (1960).

6 Joshi, U. G., and Amin, G. C., *J. Indian Chem. Soc.*, **38**, 159 (1961).

7 Patwardhan, B. H., *J. Indian Chem. Soc.*, **50**, 430 (1973).

8 Dodwadmath, R. P., and Wheeler, T. S., *Proc. Indian Acad. Sci.*, **2A**, 438 (1935).

9 Hanson, G. A., *Bull. Soc. Chim. Belges*, **67**, 91 (1958); *Chem. Abstr.*, **52**, 20111d (1958).

10 Sammour, A., Selim, M. I. B., and Abd Elhalim, M. S., *Egypt. J. Chem.*, **15**, 23 (1972); *Chem. Abstr.*, **79**, 115259q (1973).

11 Essawy, A., and Hamed, A. A., *Indian J. Chem.*, **16B**, 880 (1978).

12 Sammour, A., Raouf, A., Elkasaby, M., and Ibrahim, M. A., *Acta Chem. (Budapest)*, **78**, 399 (1973); *Chem. Abstr.*, **79**, 146100k (1973).

13 Sammour, A., Fahmy, A. F., Abd El-Rahman, S., Akhmookh, Y., and Abd El-moez, M. S., *U. A. R. J. Chem.*, **14**, 581 (1971); *Chem. Abstr.*, **79**, 115413k (1973).

14 Jahine, H., Zaher, H. A., Sayed, A. A., and Sherif, O., *Indian J. Chem.*, **11**, 1122 (1973).

15 Abdalla, M., Essawy, A., and Deeb, A., Pak, *J. Sci. Ind. Res.*, **20**, 139 (1977).

16 Toma, S., *Coll. Czech. Chem. Commun.*, **34**, 2771 (1969).

17 Xicluna, A., Guinchard, C., Robert, J. F., and Panouse, J. J., *C. R. Hebd. Seances Acad. Sci., Ser. C*, **280**, 287 (1975); *Chem. Abstr.*, **82**, 155994u (1975).

18 Kinastowski, S., Kojkowicz-Balcerek, M., Grabarkiewicz, J., and Kasprzyk, H., *Bull. Acad. Polon. Sci., Ser. Sci. Chim.*, **23**, 199 (1975); *Chem. Abstr.*, **83**, 78064g (1975).

19 Filler, R., Beaucaire, V. D., and Kang, H. H., *J. Org. Chem.*, **40**, 935 (1975).

20 Holden, N. E., and Lapworth, A., *J. Chem. Soc.*, 2368 (1931).

21 Weizmann, C., U.S. Patent 2,472,135 (1949); *Chem. Abstr.*, **43**, 6665b (1949).

22 Weizmann, C., Brit. Patent 594,182 (1947); *Chem. Abstr.*, **42**, 2987b (1948).

23 Michael, A., and Ross, J., *J. Am. Chem. Soc.*, **55**, 1632 (1933).

24 Orlov, V. D., Tishchenko, V. N., Ivanova, E. T., and Lavrushin, V. F., *Zh. Org. Khim.*, **13**, 2142 (1977); *Chem. Abstr.*, **88**, 104840g (1978).

25 Michael, A., and Ross, J., *J. Am. Chem. Soc.*, **54**, 407 (1932).

26 Connor, R., and Andrews, D. B., *J. Am. Chem. Soc.*, **56**, 2713 (1934).

27 de Benneville, P. L., Clagett, D. D., and Connor, R., *J. Org. Chem.*, **6**, 690 (1941).

28 Gupta, R. K., and George, M. V., *Tetrahedron*, **31**, 1263 (1975).

Chapter Eight
Reaction of Chalcones with Amides

UREA

Chalcone reacts, under acid catalysis, with urea[1] to yield 2-oxo-4,6-di-phenyl-1,2,3,4-tetrahydropyrimidine. A similar reaction is reported to take place with arylidene bis-ureas.[1]

THIOUREA

2-Oxo-4,6,-diphenyl-3,6-dihydro-1,3-thiazine (I) is obtained by the reaction of chalcone with thiourea in the presence of dilute sulfuric
acid.[2] The formation of I has been rationalized as follows[2]:

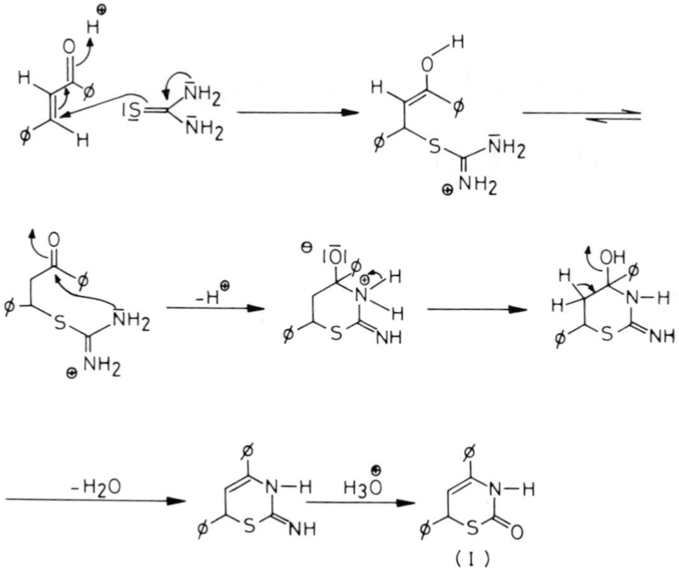

4-Chloro- (and 4-methoxy-) chalcones, however, yield their corresponding dimers[2] when they react under the above conditions.

CYANOACETAMIDE

Chalcone condenses with cyanoacetamide[3] to yield a nitrogen heterocycle (II):

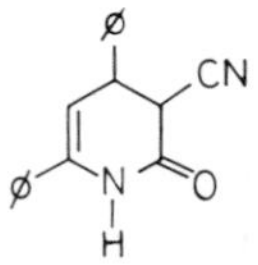

(II)

MALONAMIDE

Depending on the conditions of the experiment, different products, **III** or **IV**, are formed by the reaction of chalcones with malonamide.[4]

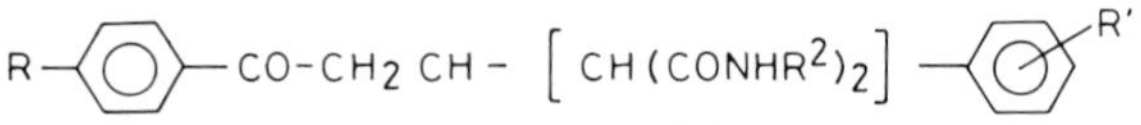

$$R-\langle\bigcirc\rangle-CO-CH_2\ CH-\left[CH(CONHR^2)_2\right]-\langle\bigcirc\rangle^{R'}$$

(III)

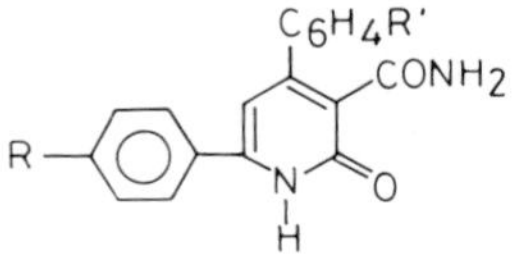

$$R-\langle\bigcirc\rangle\underset{\underset{H}{N}}{\overset{C_6H_4R'}{\underset{\|}{\longleftarrow}}}\overset{CONH_2}{\underset{O}{}}$$

(IV)

N,N-DIBROMOBENZENESULFONAMIDE[5,6]

The reaction of *N,N*-dibromobenzenesulfonamide with chalcones presents an interesting example. Thus different products[6] are formed with variously substituted chalcones:

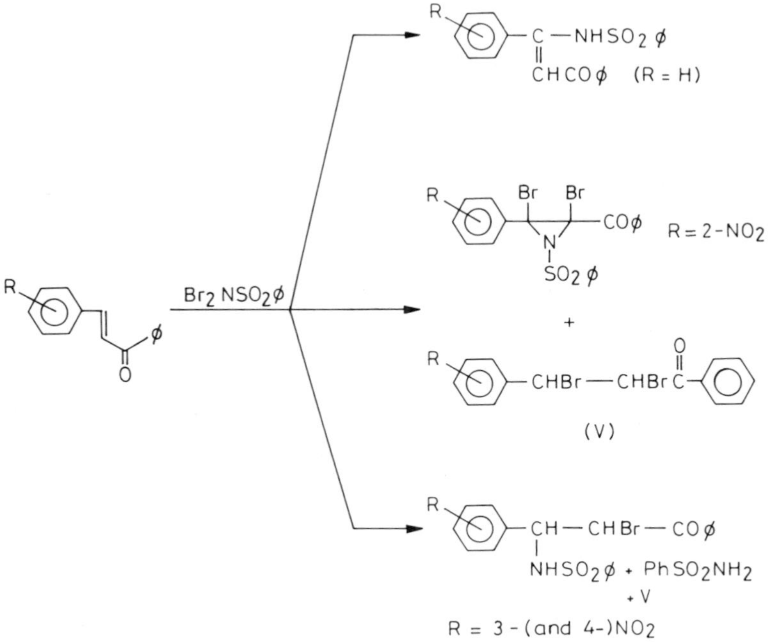

$$R-\langle\bigcirc\rangle-\underset{\underset{CHCO\phi}{\|}}{\overset{C-NHSO_2\ \phi}{}}\quad (R = H)$$

$$R-\langle\bigcirc\rangle\underset{\underset{SO_2\ \phi}{N}}{\overset{Br\quad Br}{\underset{\|}{}}}-CO\phi\quad R = 2-NO_2$$

+

$$R-\langle\bigcirc\rangle-CHBr-CHBr\ \overset{O}{\overset{\|}{C}}-\langle\bigcirc\rangle$$

(V)

$$R-\langle\bigcirc\rangle-\underset{\underset{NHSO_2\phi}{}}{\overset{CH}{}}-CHBr-CO\phi$$

$NHSO_2\phi$ + $PhSO_2NH_2$
+ V

R = 3 - (and 4-)NO_2

Br_2 NSO_2φ

$$R-\langle\bigcirc\rangle\underset{O}{\overset{}{\diagdown}}\phi$$

AROMATIC THIOAMIDES

1,3-Thiazinium salts[7] derived from chalcone have been secured by the following series of reactions:

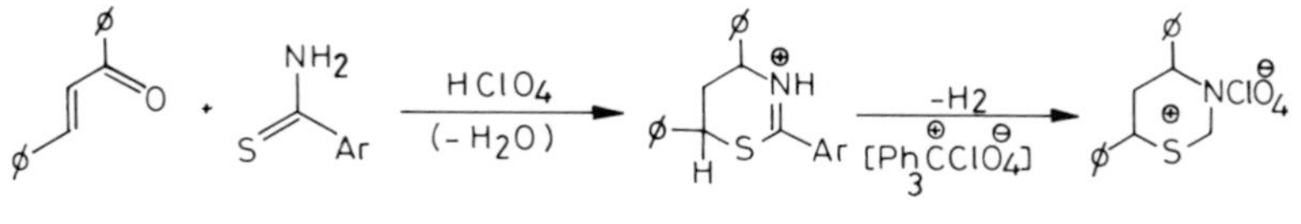

REFERENCES

1 Sedova, V. F., Samsonov, V. A., and Mamaev, V. P., *Izv. Sib. Otd. Akad. Nauk, SSSR, Ser. Khim. Nauk*, **2**, 112 (1972); *Chem. Abstr.*, **77**, 164635b (1972).

2 Dhar, D. N., and Singh, A. K., Annual Convention of Chemists Abstracts, Org-57, p. 22 (1976); *Indian J. Chem.*, **17B**, 25 (1979).

3 Carrie, R., and Rochard, J. C., *Compt. Rend.*, **257**, 2849 (1963).

4 Sammour, A., Selim, M. I. B., and Abdel Halim, M. S., *Egypt. J. Chem.*, **15**, 23 (1972); *Chem. Abstr.*, **79**, 115259q (1973).

5 Chen, C-T., *Bull. Inst. Chem., Acad. Sinica*, No. 15, 53 (1968); *Chem. Abstr.*, **70**, 77503u (1969).

6 Chen, C-T., and Chen, T-K., *Bull. Inst. Chem., Acad. Sinica*, **22**, 88 (1973); *Chem. Abstr.*, **81**, 135629e (1974).

7 Hartmann, H., *Tetrahedron Letters*, 3977 (1972).

Chapter Nine
Reaction of Chalcones with Cyanides and Isocyanates

HYDROCYANIC ACID

Hydrocyanic acid is reported to react with chalcone to yield an adduct, $\phi CH(CN)CH_2CO\phi$.[1] The adduct is transformed into acid, lactone, pyridazinone, and diol under appropriate reaction conditions.[1]

The exchange of hydrocyanic acid occurs between chalcone and acetone cyanohydrin[2] in the presence of a base. Reaction may take either of the following pathways, depending on the condition of the experiment.[2] Thus

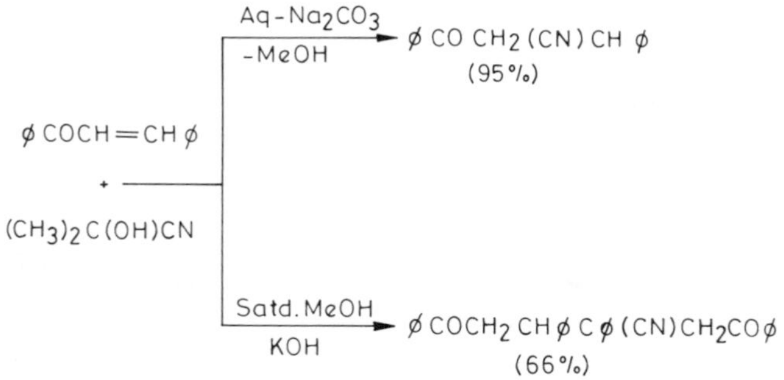

ϕ COCH=CHϕ

+

$(CH_3)_2 C(OH)CN$

$\xrightarrow[\text{-MeOH}]{\text{Aq-Na}_2\text{CO}_3}$ ϕ CO CH$_2$(CN)CH ϕ
(95%)

$\xrightarrow[\text{KOH}]{\text{Satd.MeOH}}$ ϕ COCH$_2$ CHϕ Cϕ(CN)CH$_2$COϕ
(66%)

ALIPHATIC AND ALICYCLIC NITRILES

Aliphatic nitriles, RCH_2CN, add across the carbonyl group of chalcone to give β-hydroxynitriles.[3] Thus acetonitrile[3] reacts with chalcone in the presence of lithium amide in liquid ammonia to give ϕCH=CHC(OH)(CH$_2$CN)ϕ.

BENZONITRILE AND ACETONITRILE

Chalcone behaves as a less satisfactory olefin component in the Ritter reaction. Thus benzonitrile reacted with chalcone in the presence of acid, resulting in the formation of 3-benzamido-3-phenylpropiophenone[4] in a small yield. With acetonitrile, however, the end product is an oxazine[4]:

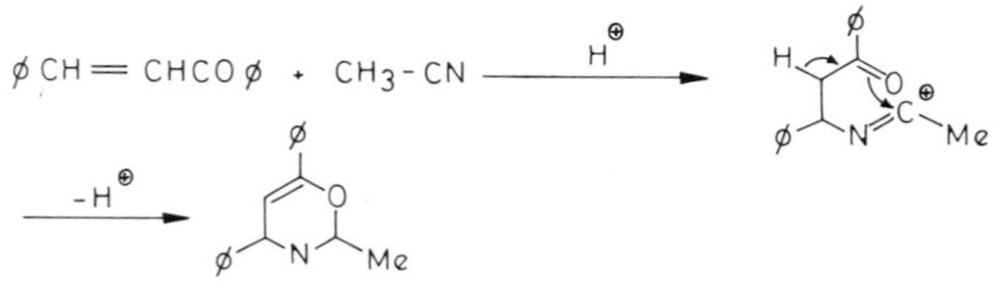

ϕ CH=CHCOϕ + CH$_3$-CN $\xrightarrow{\text{H}^\oplus}$

$\xrightarrow{-\text{H}^\oplus}$

α-DIMETHYLAMINOPHENYLACETONITRILE

Chalcone undergoes conjugate addition with α-dimethylaminopheny-lacetonitrile in the presence of potassium amide in liquid ammonia. The product α,β,δ-triphenyl-α-dimethylamine-δ-ketovaleronitrile[5] is obtained in 84% yield.

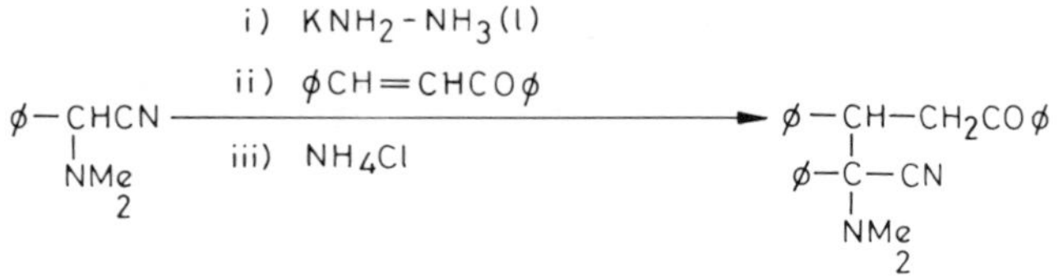

$$
\phi-\underset{\underset{2}{\overset{|}{NMe}}}{\overset{|}{CH}}CN
\quad\xrightarrow[\text{iii) } NH_4Cl]{\begin{array}{l}\text{i) } KNH_2-NH_3(l)\\ \text{ii) } \phi CH=CHCO\phi\end{array}}\quad
\phi-\underset{\phi-\underset{\underset{2}{\overset{|}{NMe}}}{\overset{|}{C}}-CN}{\overset{|}{CH}}-CH_2CO\phi
$$

ETHYL-γ-CYANO-β,γ-DIPHENYLBUTYRATE

Chalcone reacts with the title compound in the presence of sodium ethoxide in ethanol to give 3-benzoyl-4-keto-1,2,6-triphenylcyclohex-anenitrile.[6]

POTASSIUM CYANIDE

Chalcone and potassium cyanide[7] in boiling methanol yield an addition product, $\phi CH[C\phi(CN)CH_2—CO\phi]CH_2CO\phi$, according to the following scheme:

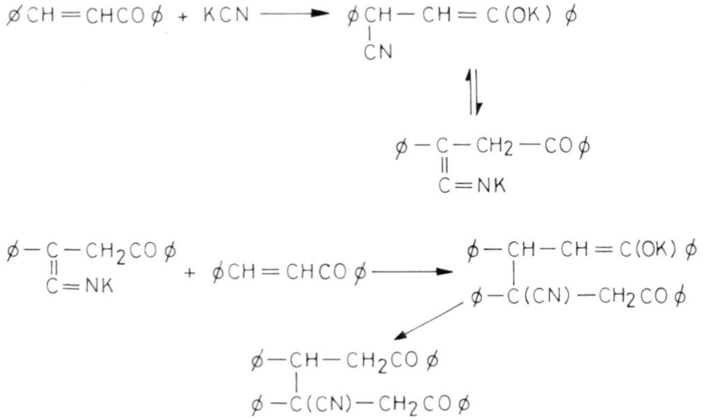

$$
\phi CH=CHCO\phi + KCN \longrightarrow \phi\underset{\overset{|}{CN}}{CH}-CH=C(OK)\phi
$$

$$
\Updownarrow
$$

$$
\phi-\underset{\overset{\|}{C}=NK}{C}-CH_2-CO\phi
$$

$$
\phi-\underset{\overset{\|}{C}=NK}{C}-CH_2CO\phi \;+\; \phi CH=CHCO\phi \longrightarrow
\begin{array}{l}\phi-\underset{\phi-\overset{|}{C}(CN)-CH_2CO\phi}{\overset{|}{CH}}-CH=C(OK)\phi\end{array}
$$

$$
\phi-\underset{\phi-\overset{|}{C}(CN)-CH_2CO\phi}{\overset{|}{CH}}-CH_2CO\phi
$$

CHLOROSULFONYL ISOCYANATE

Chlorosulfonyl isocyanate is reported to react with chalcone to yield N-chlorosulfonyl-4,6-diphenyl-2-oxo-3,4-dihydro-1,3-oxazine   (Ia, 55%); the latter compound on reduction with thiophenol–pyridine gives **Ib**.[8]

In the case of 4- (and 4'-) methoxychalcones, however, the reaction with chlorosulfonyl isocyanate leads to the formation of N-chlorosulfonylimine,[9] R'—C($=$NSO$_2$Cl)CH$=$CH—R, via the unstable 1,3-oxazetidin-2-one.

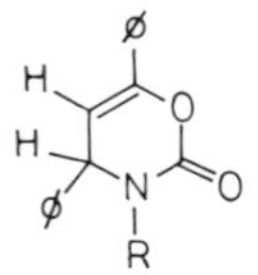

(I)   (a) R = SO$_2$Cl ;
        (b) R = H

REFERENCES

1 Davey, W., and Tivey, D. J., *J. Chem. Soc.*, 1230 (1958).

2 Betts, B. E., and Davey, W., *J. Chem. Soc.*, 4193 (1958).

3 Popandova, K. V., and Ivanov, K., *Dokl. Bolg. Akad. Nauk*, **24**, 621 (1971); *Chem. Abstr.*, **75**, 98296x (1971).

4 Scheuer, P. J., Botelho, H. C., and Pauling, C., *J. Org. Chem.*, **22**, 674 (1957).

5 Taylor, H. M., and Hauser, C. R., *J. Am. Chem. Soc.*, **82**, 1790 (1960).

6 Helmkamp, R. W., Tanghe, L. J., and Plati, J. T., *J. Am. Chem. Soc.*, **62**, 3215 (1940).

7 Michael, A., and Weiner, N., *J. Am. Chem. Soc.*, **59**, 744 (1937).

8 Dhar, D. N., Mehta, G., and Suri, S. C., *Indian J. Chem.*, **14B**, 477 (1976).

9 Dhar, D. N., and Suri, S. C., *Indian J. Chem.*, **18B**, 281 (1979).

Chapter Ten

Reaction of Chalcones with Amines

METHYLAMINE

N,N-Bis(α-phenacylbenzyl)methylamine, $[CH_3N(—CH\phi)CH_2CO\phi]_2$, results by the interaction of chalcone with methylamine[1] in ethanolic solution.

METHOXYAMINE

Methoxyamine[2] reacts with chalcone and its *p*- or *p'*-substituted analogue to give the addition product β-methoxyaminopropiophenone
and β,β'-methoxyiminobispropiophenone. In the presence of a strong
base the addition product obtained, for example, in the case of chalcone, undergoes rearrangement[2] to yield α-aminochalcone (94%):

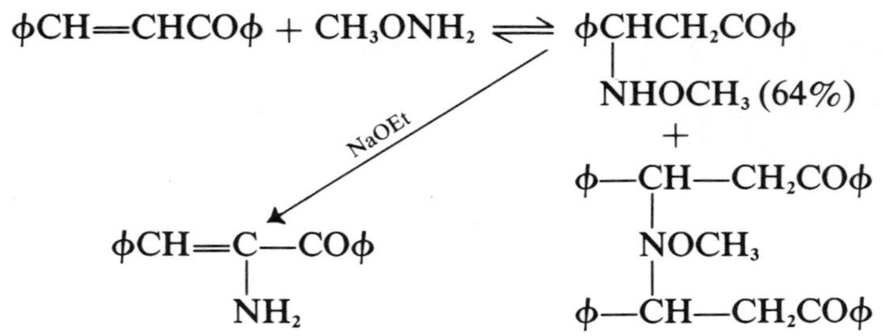

$$\phi CH=CHCO\phi + CH_3ONH_2 \rightleftharpoons \phi CHCH_2CO\phi$$

$$\underset{NHOCH_3}{|} (64\%)$$

$$+$$

$$\phi\text{---}CH\text{---}CH_2CO\phi$$

$$\underset{NOCH_3}{|}$$

NaOEt

$$\phi CH=C\text{---}CO\phi$$

$$\underset{NH_2}{|}$$

$$\phi\text{---}CH\text{---}CH_2CO\phi$$

Aziridines have been prepared from chalcones in the following
way[3]:

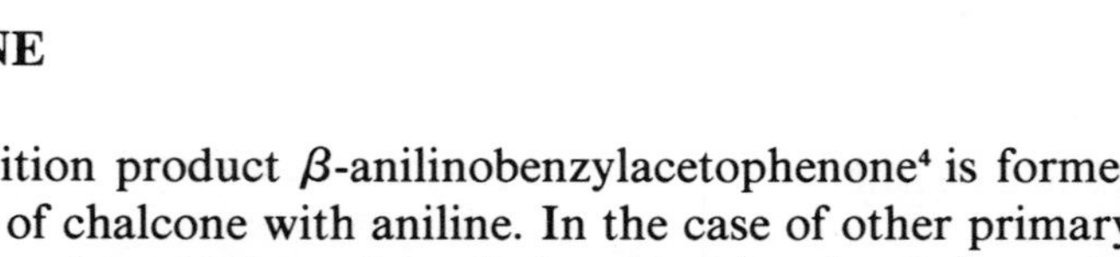

$$MeO-C_6H_4COCH=CH\phi \xrightarrow{MeONH_2} MeO-C_6H_4COCH_2-$$

$$CH(NHOMe)\phi \xrightarrow{NaOMe/MeOH} MeO-C_6H_4CO-CH\overset{\displaystyle N}{\underset{H}{\diagdown\diagup}}CH-\phi$$

ANILINE

The addition product β-anilinobenzylacetophenone[4] is formed by the
reaction of chalcone with aniline. In the case of other primary aromatic amines the addition of the hydrochloride salt of the amine is preferred,[5] and the yields of products range from 27 to 55%.

Nitromethylchalcones have been reported to react with *p*-nitro-
N,N-dimethylaniline (NDA), illustrating the reaction capacity of the
methyl group.[6]

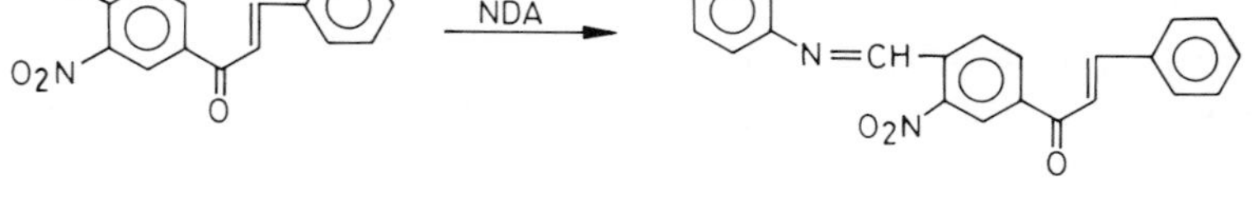

1-CYANO-2-METHYL-2-AMINOETHYLENE

3-Cyano-4-phenylpyridines have been secured by the reaction of 1-cyano-2-methyl-2-aminoethylene with chalcones, which have been subsequently transformed into 2-azafluorenones.[7]

α-AMINOBENZENETHIOL

Two products are formed in the reaction of α-aminobenzenethiol with chalcone; β-phenyl-β-(o-aminophenylmercapto)propiophenone and its cyclized product, 2,4-diphenyl-6,7-benzo-1-thia-5-aza-4,6-cyclohepta-diene[8]:

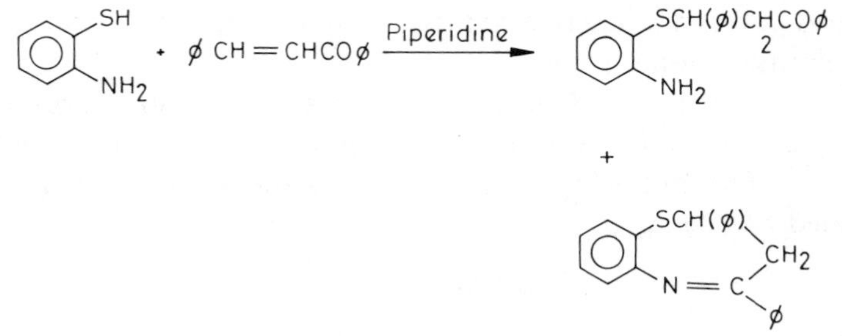

PYRROLIDINE

The reaction of *cis*- and *trans*-chalcones with pyrrolidine has been investigated.[9] It has been shown that *cis*-chalcone changes to the corresponding *trans* isomer before yielding the Michael adduct[9]:

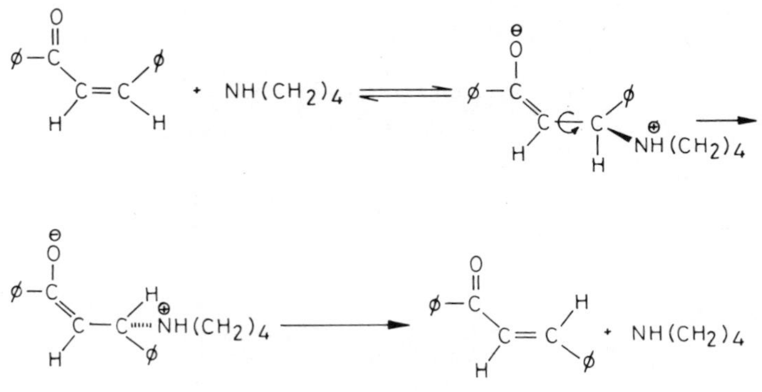

IODINE–AMINE COMPLEXES

Ethyleneimine ketones[10] are formed by the action of iodine complexes of ammonia or primary amines with chalcones. α,β-Diaminobenzylacetophenone, however, is produced when chalcone reacts with a secondary amine–iodine complex.

It has been observed that in the preparation of ethyleneimine ketone the nature of the solvent has a decisive influence on the configuration of the end product.[11] For example,[11] the reaction of *trans*-chalcone with a cyclohexylamine–iodine complex in methanolic solution yields 88% of *trans*-1-cyclohexyl-2-phenyl-3-benzoylethyleneimine, as compared to 61–73% in benzene. On the other hand, substitution of methanol for benzene in the reaction of cyclohexylamine with *trans*-α-bromochalcone greatly increased the proportion of *cis*-ethyleneimine ketone.

ortho-Substituted chalcones react with amines[12,13] in the presence of iodine to give different products, depending on the nature of the 2′-substituent. The following table illustrates the various products that are formed:

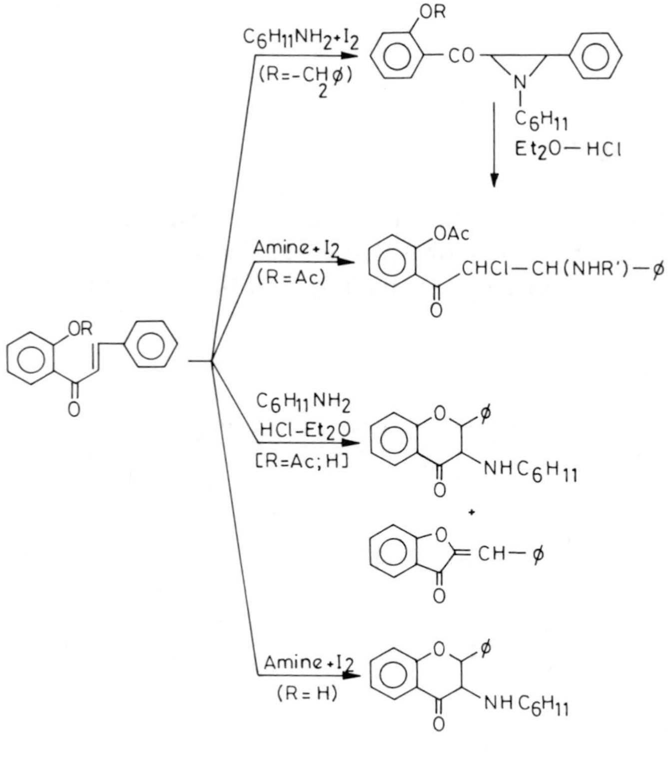

TRIAMINOGUANIDINE SALT

The reaction of chalcone with triaminoguanidine[14] is slow and yields the corresponding hydrazone (75%) of the type $(RR'C{=}NNH)_2$ $C{=}N{-}N{=}CRR'$, where R=phenyl and R'=styryl groups.

AMIDINES[15,16]

Amidine hydrochloride is reported to react with chalcone to yield the cycloaddition product, 2,4,6-triphenylpyrimidine, in 85% yield:

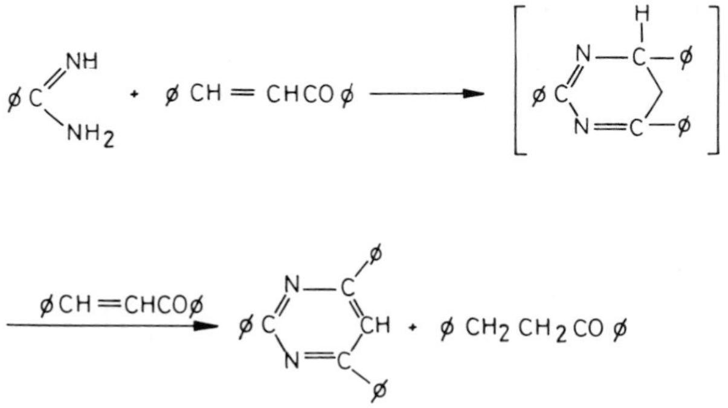

In this reaction a part of the chalcone is reduced to the saturated ketone.

REFERENCES

1   Cromwell, N. H., and Caughlan, J. A., *J. Am. Chem. Soc.*, **67**, 2235 (1945).

2   Blatt, A. H., *J. Am. Chem. Soc.*, **61**, 3494 (1939).

3   Drefahl, G., Ponsold, K., and Schoenecker, B., *Chem. Ber.*, **97**, 2014 (1964).

4   Cromwell, N. H., Wiles, Q. T., and Schroeder, O. C., *J. Am. Chem. Soc.*, **64**, 2432 (1942).

5   Kozlov, N. S., and Shur, I. A., *Zh. Obshch. Khim.*, **30**, 2746 (1960).

6   Chardonnens, L., and Venetz, J., *Helv. Chim. Acta*, **22**, 1278 (1939).

7   Chatterjee, J. N., and Prasad, K., *J. Sci. Ind. Research (India)*, **14B**, 383 (1955).

8   Stephens, W. D., and Field, L., *J. Org. Chem.*, **24**, 1576 (1959).

9   Menger, F. M., and Smith, J. H., *J. Am. Chem. Soc.*, **91**, 4211 (1969).

10   Southwick, P. L., and Christman, D. R., *J. Am. Chem. Soc.*, **74**, 1886 (1952).

11   Southwick, P. L., and Shozda, R. J., *J. Am. Chem. Soc.*, **82**, 2888 (1960).
12   Bognar, R., Litkei, G., and Szigeti, P., *Acta Chim.* (*Budapest*), **68**, 421 (1971); *Chem. Abstr.*, **75**, 35620c (1971).
13   Litkei, G., Bognar, R., Szigeti, P., and Trapp, V., *Acta Chim.* (*Budapest*), **73**, 71 (1972); *Chem. Abstr.*, **77**, 88219b (1972).
14   Scott, F. L., Cashman, M., and Reilly, J., *J. Am. Chem. Soc.*, **75**, 1510 (1953).
15   Dodson, R. M., and Seyler, J. K., *J. Org. Chem.*, **16**, 461 (1951).
16   Vais, A. L., Shyrina, V. M., and Mamaev, V. P., *Izv. Sib. Otd. Akad. Nauk SSSR, Ser. Khim. Nauk*, (6), 144 (1975); *Chem. Abstr.*, **84**, 105528r (1976).

Chapter Eleven

Reactions of Chalcone with Organometallic and Organomagnesium Compounds

REACTION WITH ORGANOMETALLIC COMPOUNDS

Lithium Acetylide

The reaction of chalcone with lithium acetylide in liquid ammonia yields the corresponding ethynylcarbinol[1] (1:2 adduct) in a good yield (81%).

Phenylalkali Metals

Phenyllithium, phenylsodium, and phenylpotassium react with chalcone to yield mainly the 1,2-addition products[2] (60–75%):

$$\text{Ph}-\underset{\underset{O}{\|}}{C}-\text{CH}=\text{CH}-\text{Ph} \xrightarrow{\text{PhM}} \text{Ph}-\underset{\underset{OM}{|}}{\overset{\overset{Ph}{|}}{C}}-\text{CH}=\text{CH}-\text{Ph}$$

(M = Li, Na, or K)

The mode of addition of the organometallic compound to chalcone depends upon the reactivity of the former.[3] In these cases, the addition may occur across the double bond and/or the carbonyl group of the chalcone molecule. The following examples are illustrative:

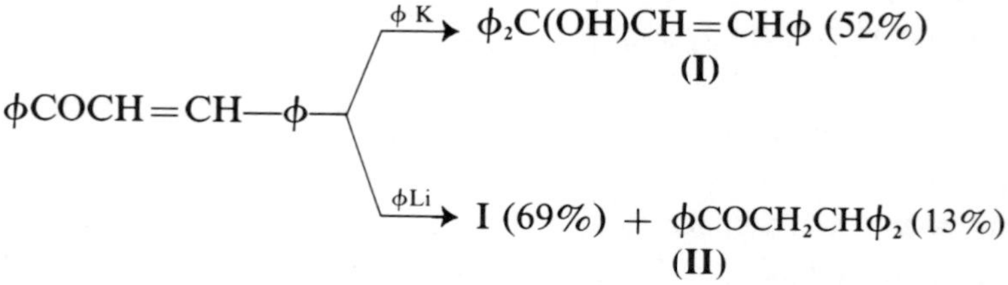

$$\phi\text{COCH}=\text{CH}-\phi \left\langle \begin{array}{l} \xrightarrow{\phi\text{K}} \phi_2\text{C(OH)CH}=\text{CH}\phi \ (52\%) \quad \textbf{(I)} \\[2em] \xrightarrow{\phi\text{Li}} \textbf{I} \ (69\%) + \phi\text{COCH}_2\text{CH}\phi_2 \ (13\%) \quad \textbf{(II)} \end{array} \right.$$

Reaction with excess phenyllithium is reported to yield 1,1,2,3,3-pentaphenylpropanol besides the diphenylstyrylcarbinol (I).[4]

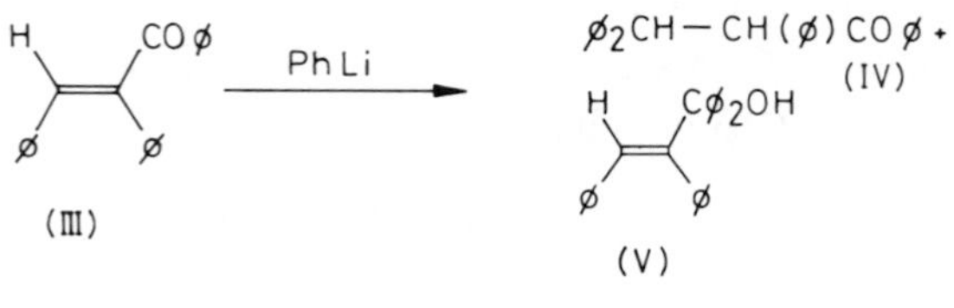

Phenyllithium is also reported to add to *trans*-α-phenylchalcone (**III**), yielding 33% of the 1,4-addition product (**IV**) and 48% of the 1,2-addition product[5]:

Low-temperature reaction of chalcone with phenyllithium, followed by decomposition of the reaction mixture with water, gives two products,[6] $PhCH=CHCPh_2OH$ (84%) and $Ph_2CH—CH_2COPh$ (16%), respectively.

Benzhydrylsodium

Chalcone reacts with benzhydrylsodium[7] according to the following reaction:

$$PhCOCH=CHPh \xrightarrow{Ph_2CHNa} PhCOCH_2—\underset{\underset{CHPh_2}{|}}{CH}—Ph$$

Arylcopper

Chalcone reacts with arylcopper[8] in the presence of aryl iodide under Ullmann conditions to give, among other products, β-arylchalcone in small amounts.

Diphenylberyllium

Diphenylberyllium[3] reacts with chalcone to give $PhCOCH_2CHPh_2$ in a high yield (~90%).

Diphenylmercury

Diphenylmercury[9] reacts with chalcone in the manner indicated:

$$PhCOCH=CHPh + Ph_2Hg \xrightarrow{150°} PhHgC_6H_4CH=CHCOPh$$
$$(76\%)$$

Triphenylaluminum, Triphenylindium, and Triphenylthallium

ββ'-Diphenylpropiophenone is obtained when chalcone reacts with triphenylaluminum[10] or triphenylindium.[11] Under the same reaction conditions triphenylthallium[12] yields an additional product, β-phenyl-γ-benzhydrylbutyrophenone (30%).

Triphenylmethylsodium

The title reagent is reported to react with chalcone to yield 2,3,3,3-tetraphenylpropyl phenyl ketone (25–35%).[4]

Triphenylcadmium–Lithium

Chalcone and triphenylcadmium–lithium[13] react to give, after hydrolysis, three products, diphenylpropiophenone (21%), diphenylstyrylcarbinol (6%), and β-phenyl-γ-benzoyl-γ-benzhydrylbutyrophenone (1.8%).

Triphenyltin–Lithium

Tetraphenyltin[14] (7.8%) is produced in the reaction of chalcone with triphenyltin–lithium.

Tetraphenylboro-Copper

The pyridine salt of tetraphenylboro-Copper[15] and chalcone after reaction furnish $\beta\beta$-diphenylpropiophenone (29%).

REACTION WITH ORGANOMAGNESIUM COMPOUNDS

Methylmagnesium Bromide

Chalcone reacts with an excess of methylmagnesium bromide to yield the 1,4-addition product, β-phenylbutyrophenone[16,17] or the bimolecular compound.[17] Sometimes a secondary product, 1,3,5-triphenyl-4-benzoylhex-1,3-diene, is obtained.

Ethylmagnesium Bromide

The products formed in the reaction of chalcone with ethylmagnesium bromide correspond to 1,2- and 1,4-addition.

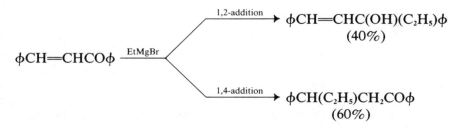

ϕCH=CHCOϕ $\xrightarrow{\text{EtMgBr}}$

1,2-addition → ϕCH=CHC(OH)(C$_2$H$_5$)ϕ
(40%)

1,4-addition → ϕCH(C$_2$H$_5$)CH$_2$COϕ
(60%)

The effect of some metallic chlorides on the reaction of chalcone with Grignard reagents is reported.[16] For the formation of reduction dimers cobalt chloride acts as a powerful catalyst compared to cuprous and ferric chlorides.

n-Butylmagnesium Bromide

The relative activities of chalcones, carrying α- or β-methyl substituents toward n-butylmagnesium bromide, have been studied.[18] β-Methyl compounds are reported to be more reactive in comparison to their α isomers.[18]

Phenylmagnesium Bromide

The reaction of the *trans*-chalcone with two equivalents of phenylmagnesium bromide yields two products, 3,3-diphenylpropiophenone[19,20] (83%) and *trans*-1,1,3-triphenyl-2-propen-1-ol (6%), respectively. However, a different product,[21] 1,1,3,3-tetraphenyl-1-propanol, results by varying the experimental conditions in the above reaction. Variation in the manner of workup of the Grignard reaction mixture, namely, the treatment with dry ice, is reported to yield 2-benzhydryl-3-hydroxy-3,3-diphenylpropionic acid.[21,22]

The mode of addition of phenylmagnesium bromide to chalcone depends on the chemical nature of the substituents.[23] Thus the principal product of the reaction with 4-carboethoxychalcone is a saturated ketone, while in the case of 4-dimethylaminochalcone the product is an α,β-unsaturated alcohol.[23]

Organomagnesium Compounds

The Grignard reaction of *trans*-chalcone with Me_2CHCH_2-$COCHXCHMe$ (X = Cl, Br) and $[Me_2CHCH_2COCH(CHMe_2)]_2Mg$ is reported to give two diastereoisomeric diketones, **VI** and **VII**[24]:

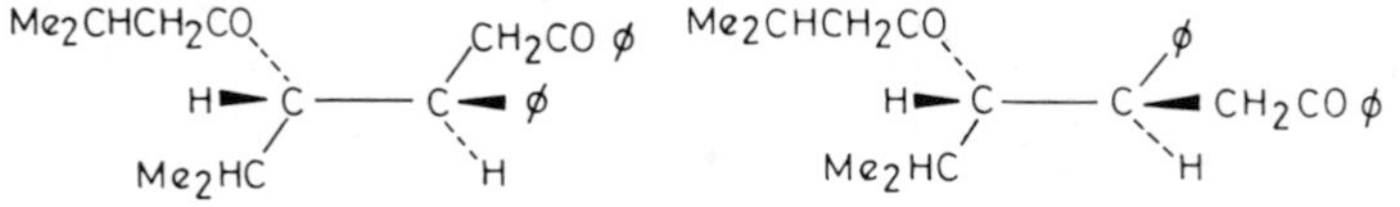

$$(VI) \qquad\qquad (VII)$$

The reaction of various magnesium carbonyls with *trans*-chalcone has been studied.[25] Whether or not the addition occurs at the carbonyl or olefinic sites of chalcone depends on the nature of the added nucleophile.

Aminomagnesium Compounds

Dypnopinacols (50–70%) have been secured by the interaction of chalcones with dypnones under the influence of an aminomagnesium compound.[26]

REFERENCES

1   Oroshnik, W., and Mebane, A. D., *J. Am. Chem. Soc.*, **71**, 2062 (1949).

2   O'Sullivan, W. I., Swamer, F. W., Humphlett, W. J., and Hauser, C. R., *J. Org. Chem.*, **26**, 2306 (1960).

3   Gilman, H., and Kirby, R. H., *J. Am. Chem. Soc.*, **63**, 2046 (1941).

4   Michael, A., and Saffer, C. M., Jr., *J. Org. Chem.*, **8**, 60 (1943).

5   Lutz, R. E., and Rinker, E. H., Jr., *J. Am. Chem. Soc.*, **77**, 366 (1955).

6   Luttringhaus, A., and Scholtis, K., *Annalen*, **557**, 70 (1945).

7   Bergmann, E., *J. Chem. Soc.*, 412 (1936).

8   Treblow, M. D., *Diss. Abstr., Int. B*, **34**, 148 (1973).

9   Koton, M. M., and Zorina, T. M., *Zh. Obsch. Khim.*, **19**, 1137 (1949).

10   Wittig, G., and Bub, O., *Annalen*, **566**, 113 (1950).

11   Gilman, H., and Jones, R. G., *J. Am. Chem. Soc.*, **62**, 2353 (1940).

12   Gilman, H., and Jones, R. G., *J. Am. Chem. Soc.*, **61**, 1513 (1939).

13   Wittig, G., Meyer, F. J., and Lange, G., *Annalen*, **571**, 167 (1951).

14   Gilman, H., and Rosenberg, S. D., *J. Org. Chem.*, **18**, 1554 (1953).

15 Sazonova, V. A., and Nazarova, I. I., *Zh. Obshch. Khim.*, **26**, 3440 (1956).

16 Kharasch, M. S., and Sayles, D. C., *J. Am. Chem. Soc.*, **64**, 2972 (1942).

17 Filler, R., Beaucaire, V. D., and Kang, H. H., *J. Org. Chem.*, **40**, 935 (1975).

18 Tsuruta, T., and Yasuda, Y., *J. Macromol. Sci. Chem.*, **2**, 939 (1968).

19 Dupre, R. A., *Diss. Abstr., B*, **30**, 556 (1969).

20 Sammour, A., and Elkasaby, M., *U.A.R. J. Chem. Soc.*, **13**, 409 (1971); Chem. Abstr., **77**, 101261f (1972).

21 Fuson, R. C., San, T., and Diekmann, J., *J. Org. Chem.*, **27**, 1221 (1962).

22 Diekmann, J., *Diss. Abstr.*, **20**, 4518 (1960).

23 MacLean, I. S., and Widdows, S. T., *J. Chem. Soc.*, **105**, 2169 (1914).

24 Gorrichon-Guigon, L., Maroni-Barnaud, Y., Maroni, P., and Bastide, J. D., *Bull. Soc. Chim. France* (1–2, Pt. 2) 291 (1975).

25 Bertrand, J., Gorrichon-Guigon, L., Koudsi, Y., Perry, M., and Maroni-Barnaud, Y., *C. R. Acad. Sci., Ser. B*, **277**, 723 (1973).

26 Ivanov, D., and Ivanov, C., *Chem. Ber.*, **76B**, 988 (1943).

Chapter Twelve

Reaction of Chalcones with Halogens, Pseudohalogens, and Interhalogen Compounds

CHLORINE

Chalcone reacts with chlorine to yield two stereoisomeric chalcone α,β-dichlorides.[1] The dichlorides,[2,3] under appropriate reaction conditions, are convertible to α-chlorochalcone (80%).

BROMINE

A large number of chalcone α,β-dibromides have been prepared.[3-13] Bromination is effected either by the reaction of bromine[3-9,11-13] (in carbon disulfide or acetic acid), tetrabromo-o-benzoquinone,[10] 1-bromo-2-methyl-2-imidazoline hydrobromide,[14] or 1,3-dibromo-5,5'-dimethylhydantoin[15] on chalcones. The reaction with bromine is sometimes attended with nuclear bromination[11-13] (ring A and/or ring B of chalcone), depending on the reaction conditions employed.

Kinetics of Bromine Addition

The kinetics of bromine addition to chalcone and its derivatives has been investigated,[16,17] and a two-step mechanism of this reaction has been proposed[18]:

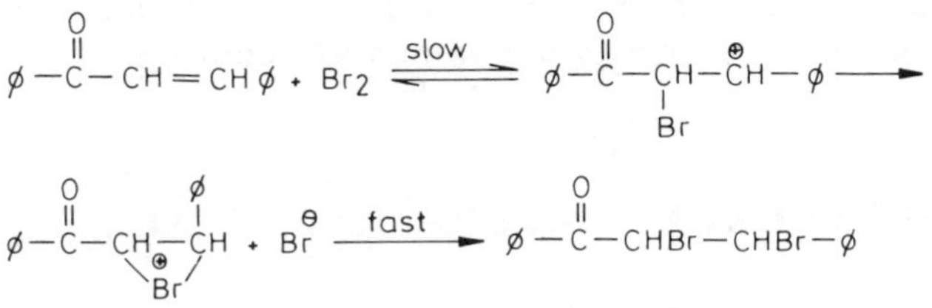

Debromination of Chalcone Dibromides

Chalcone α,β-dibromides on treatment with potassium iodide in acetone are reported to undergo debromination to yield the respective chalcones.[3,7] Debromination has been achieved by the reaction of chalcone dibromide with stannous chloride in various solvents, viz., acetone, tetrahydrofuran, and dimethylformamide.[19]

In a methanolic solution chalcone reacts with bromine (at 25–30°) to give a mixture of two products,[20,21] α-bromo-β-methoxy-β-phenylpropiophenone (23%) and chalcone α,β-dibromide (5%). The yield of the former product is raised if the reaction is conducted at higher temperature (65°). Likewise α-chloro-β-methoxy-β-phenylpropiophenone[21] has been secured in 56% yield by conducting the reaction of chalcone with chlorine (in methanol) at 25–30°. The mechanism of the reaction has been postulated as follows[21]:

$$X_2 + CH_3OH \rightleftharpoons CH_3OX + HX$$

X = Cl, Br

$$\emptyset—CO\ CH\!\!=\!\!CH—\emptyset + CH_3OX \longrightarrow \emptyset COCHX—CH(OCH_3)—\emptyset$$

IODINE MONOCHLORIDE

Treatment of chiral crystals of 4,4'-dimethylchalcone with chlorine, bromine, or iodine monochloride is reported to yield optically active dihalides.[22] The reaction of iodine chloride and 2'-hydroxychalcone has been shown to yield products depending on the nature of substituent(s).[23] Thus iodine chloride reacts with 2'-hydroxy-3',5'-dimethyl-4-methoxychalcone to yield 3-iodo-6,8-dimethylflavanone. Chalcone lacking in a strong electron donor substituent (e.g., methoxy), however, undergoes addition reaction with iodine chloride. Nevertheless, nuclear haloganation, in preference to addition reaction, is reported to occur in one case, 2'-hydroxy-4',6'-dimethoxychalcone.

BROMINE FLUORIDE AND BROMINE CHLORIDE

Bromine fluoride[24] and bromine chloride[25] are reported to add to the olefinic double bond of the chalcone molecule, thus

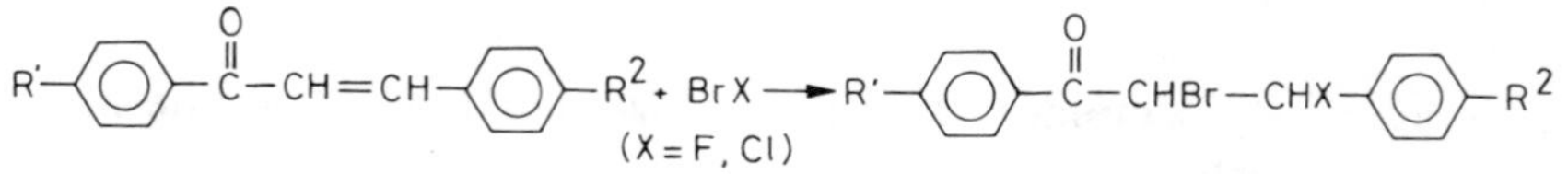

CHLORINE THIOCYANATE, THIOCYANOGEN, AND IODINE THIOCYANATE

Chlorine thiocyanate, thiocyanogen, and iodine thiocyanate[26] add across the double bond of chalcone molecule. From the addition products obtained in the first two cases α-thiocyanatochalcones have been secured by dehydrohalogenation.

REFERENCES

1 Abell, R. D., and Siddall, W., *J. Chem. Soc.*, 2804 (1953).

2 Auwers, K. V., and Hugel, R., *J. Prakt. Chem.*, **143**, 157 (1935).

3 Dodwadmath, R. P., and Wheeler, T. S., *Proc. Indian Acad. Sci.*, **2A**, 438 (1935).

4 Abell, R. D., *J. Chem. Soc.*, **101**, 998 (1912).

5 Allen, C. F. H., Normington, J. B., and Wilson, C. V., *Can. J. Res.*, **11**, 382 (1934).

6 Filler, R., Beaucaire, V. D., and Kang, H. H ., *J. Org. Chem.*, **40**, 935 (1975).

7 Merchant, J. R., and Choughuley, A. S. U., *Cur. Sci. (India)*, **29**, 93 (1960).

8 Kanthi, R. B., and Nargund, K. S., *J. Karnatak Univ.*, **2**, 8 (1957); *Chem. Abstr.*, **53**, 8067b (1959).

9 Ambekar, S. Y., Jolad, S. D., and Rajagopal, S., *Monatsh. Chem.*, **92**, 1303 (1961).

10 Latif, N., and El-Bayouki, K., *Chem. Ind. (London)*, 316 (1975).

11 Chhaya, G. S., Trivedi, P. L., and Jadhav, G. V., *J. Univ. Bombay*, **26A**, Part 3, 16 (1957); *Chem. Abstr.*, **53**, 10125b (1959).

12 Chhaya, G. S., Trivedi, P. L., and Jadhav, G. V., *J. Univ. Bombay*, **27A**, Part 3, 26 (1958); *Chem. Abstr.*, **54**, 8807e (1960).

13 Wagh, S. P., and Jadhav, G. V., *J. Univ. Bombay*, **27A**, Part 3, 1 (1959); *Chem. Abstr.*, **54**, 6707f (1960).

14 Tsuchiya, C., *Nippon Kagaku Zasshi*, **82**, 723 (1961); *Chem. Abstr.*, **59**, 599d (1963).

15 Giovambattista, N., Rabassa, R. J., and Orazi, O. O., *An. Asoc. Quim. Argent.*, **39**, No. 191, 31 (1951); *Chem. Abstr.*, **47**, 2732a (1953).

16 Mohapatra, S. N., Mohapatra, P. K., Nayak, P. L., and Rout, M. K., *J. Indian Chem. Soc.*, **49**, 135 (1972).

17 Rothbaum, H. P., Ting, I., and Robertson, P. W., *J. Chem. Soc.*, 980 (1948).

18 Rout, N., Nayak, P. L., and Rout, M. K., *J. Indian Chem. Soc.*, **47**, 217 (1970).

19 Ghiya, B. J., *Indian J. Chem.*, **9**, 502 (1971).

20 Conant, J. B., and Jackson, E. L., *J. Am. Chem. Soc.*, **46**, 1727 (1924).

21 Jackson, E. L., *J. Am. Chem. Soc.*, **48**, 2166 (1926).

22 Green, B. S., and Heller, L., *Science*, 185 (4150), 525 (1974).

23 Weber, F. G., and Brosche, K., *Z. Chem.*, **9**, 341 (1969).

24 Weber, F. G., Giese, H., and Westphal, G., *Z. Chem.*, **15**, 475 (1975).

25 Weber, F. G., and Reimann, E., *Z. Chem.*, **12**, 176 (1972).

26 Weber, F. G., Holzenger, A., Westphal, G., and Puseh, U., *Pharmazie*, **30**, 800 (1975); *Chem. Abstr.*, **84**, 89760c (1976).

Chapter Thirteen

Cyclization Reactions of Substituted Chalcones

FORMATION OF FLAVANONES

The conversion of substituted 2'-hydroxychalcones into their corresponding flavanones is usually effected under the influence of mineral acids.[1-16] The optimum time of reaction in the presence of phosphoric acid has been reported to be in the range of 20–30 hours.[8] The reaction time, however, varies with the nature of the alcohol used as a solvent.[8] Higher alcohols are not suitable, since these promote side reactions. The yield of flavanones is by and large independent of the concentration of phosphoric acid.[8]

    The preparation of d-7-hydroxyflavanone has been achieved by heating 2',4'-dihydroxychalcone with d-camphorsulfonic acid.[1] The temperature and duration of heating are reported to be important factors for the production of maximum optical activity.[1]

Some of the hydroxynitrochalcones (bearing a 2'-hydroxyl) have been cyclized to their corresponding flavanones. The chelation of the nitro group with 2'-hydroxyl is an important parameter in determining the stability of chalcone in its conversion to flavanone.[7]

Often dilute alkali has been used to bring about the cyclization of polyhydroxychalcones (e.g., butein) into their corresponding flavanones.[17,18] The synthesis of several flavanones and 3-benzylidene flavanones, using appropriately substituted chalcones as the starting materials, are reported in the literature.[19-32,37,38]

The quantitative influence of substituent effects on the cyclization kinetics of 2'-hydroxychalcone into flavanone has been studied.[33] Reaction mechanisms have been proposed[34-36] for this type of transformation.

α- or β-CYCLIZATION[39]

The α- or β-cyclization of chalcone to yield aurone or flavanone depends on the nature of the α-substituent (with respect to the carbonyl group) and pH of the reaction medium. The parameters regulating the β-cyclization include the stability of the resulting flavanones, the acidity of the 3-protons, and the presence of a 5-hydroxylic function.

FORMATION OF BENZYLIDENE COUMARANONES, CHROMONES, AND CHROMANONES

Under appropriate reaction conditions hydroxychalcones have been transformed into benzylidene coumaranones[40-42] and hydroxychromones.[43,44]

By carrying out the AFO reaction on 2'-hydroxychalcones containing an α-phenyl substituent the synthesis of chromanones is achieved.[45]

FORMATION OF FLAVONOLS, [11,12,25,26,40-42,46-51,67,70,89,90] AURONES, AND DIHYDROFLAVONOLS[68]

2'-Hydroxychalcones on oxidation with alkaline hydrogen peroxide give flavonols.[11-13,40-42,46-54] This reaction is often referred to as the Algar–Flynn–Oyamada (AFO) reaction. However, if there is a methoxy[55] or methyl substituent at the 6'-position in chalcone,[55-57] aurones are formed, provided there is no hydroxylic function present at the 2- or 4-positions[58,59] and that the reaction is carried out at room temperature. An aurone derivative (**I**) isolated in the AFO reaction of 2'-hydroxy-α,4',6'-trimethoxychalcone has been shown to have an *erythro* configuration.[60]

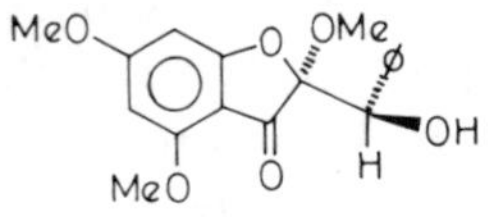

(I)

In some cases, however, mixture of aurone and flavonol are obtained.[61] For example,

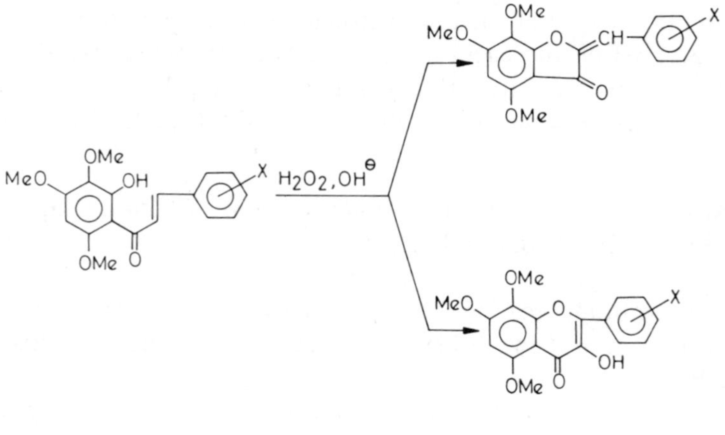

Similar results have been reported for 2',4'-dihydroxy-3'-nitrochalcone (**II**) and 2',6'-dihydroxy-3'-nitrochalcone (**III**) when these are treated with pyridine and alkaline hydrogen peroxide.[62] Thus, under these reaction conditions **II** yields 7-hydroxy-8-nitroflavonol[62] and **III** gives the corresponding aurone.[62] The chalcone bearing a nitro substituent (X = NO$_2$), however, fails to undergo the AFO reaction.[61]

Two additional products, **IV** and **V**, are reported to be formed in the AFO reaction of chalcone.[63,64] The rationale for the formation of these compounds is described,[65] and a detailed mechanism of AFO reaction has been published.[66]

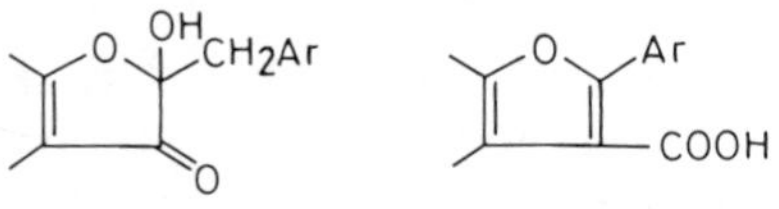

(IV)   (V)

Flavanones[24,25] can also be converted into flavonols with alkaline hydrogen peroxide. In this case the reaction is reported to proceed through the intermediacy of 2'-hydroxychalcone and dihydroflavonol.[24-25]

Hydroxyflavonols[67] have been secured by the rearrangement of α-nitrochalcone epoxides.

The geometric isomers of dihydroxyflavonols have been prepared.[68] The following sequence of steps were employed, for example, in the synthesis of *cis*- and *trans*-3',4',7-trimethoxydihydroflavonols.

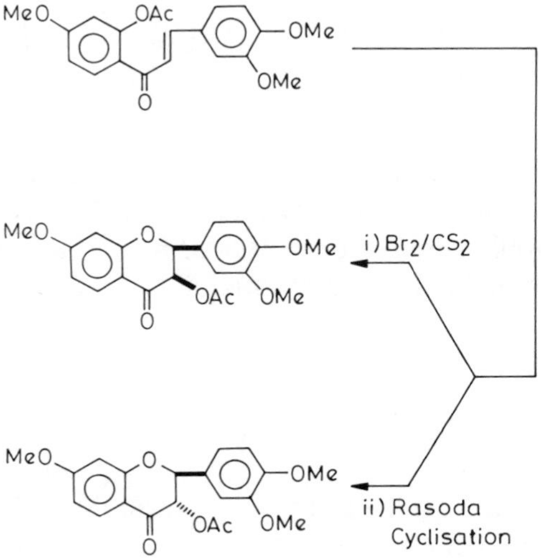

i) Br$_2$/CS$_2$

ii) Rasoda Cyclisation

FORMATION OF FLAVONES

Flavones have been prepared from appropriately substituted chalcones either by their oxidation with selenium dioxide,[13,23,31,69–81] or by heating with palladium black.[82,83] For example, cirsimaritin, a flavone derivative, has been secured by the reaction of 2'-hydroxy-4-benzoyloxy-4',5',6'-trimethoxychalcone with selenium dioxide, followed by debenzoylation and partial demethylation.[71] The syntheses of naturally occurring flavones,[74] cirsiliol and cirsilineol, are based on the aforesaid reaction. 7-Hydroxyflavone, on the other hand, has been secured in 35% yield by heating (at 220°) 2',4'-dihydroxychalcone with palladium.[82]

An alternative method[84] of preparing flavones in high yield, is the reaction of 2'-hydroxychalcone (sodium salt) with lithium chloropalladite (LCP). The following mechanism has been postulated for the reaction:

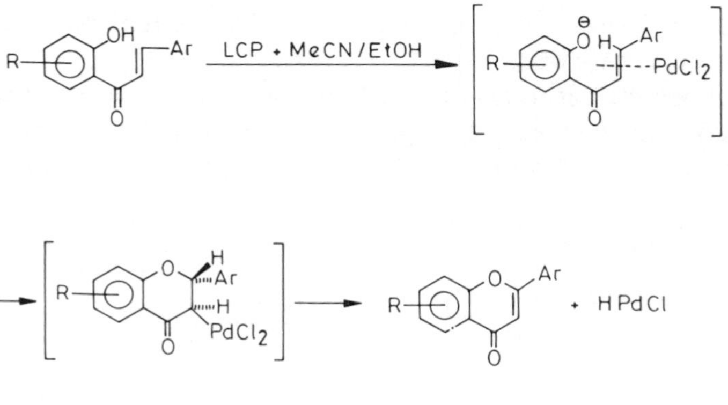

Several 3-bromoflavones are reported to have been obtained as a mixture of *axial* and *equitorial* isomers from 2'-hydroxy-4-methoxy-chalcones by initial bromination, followed by iodine-catalyzed cyclization of the resulting chalcone dibromide.[91]

FORMATION OF ISOFLAVONES

Two methods[85–87] are available for the synthesis of isoflavones from chalcones. The first method involves the preparation of an epoxide, followed by BF_3-catalyzed rearrangement and cyclization. The preparation of 3',7-dihydroxy-4'-methoxyisoflavone is illustrative[85]:

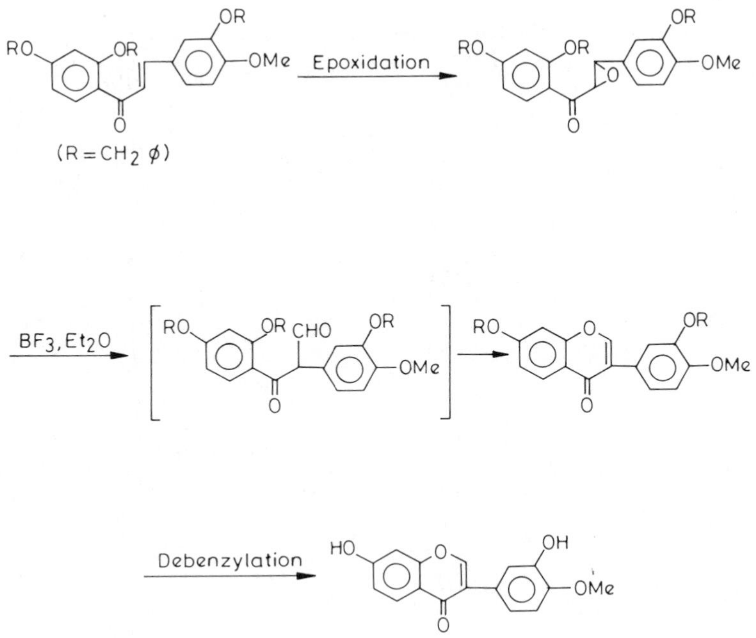

$(R = CH_2 \phi)$

The second method involves the oxidative rearrangement of an appropriately substituted chalcone with thallic nitrate.[86] The naturally occurring 2',3',4',6,7-pentamethoxy- and 2',4',5',6,7-pentamethoxyisoflavones have been secured by this method.[86]

CYCLIZATION OF 2'-HYDROXY-2-NITRO-5'-METHYLCHALCONE[88]

The title compound undergoes cyclization in the presence of caustic alkali to yield 10-hydroxy-2-methyl-11H[1]-benzopyrano[3,2-b]indole-

11-one **(VI)**. Treatment of the latter with sulfur dioxide in methanol yields the following compound **(VII)**:

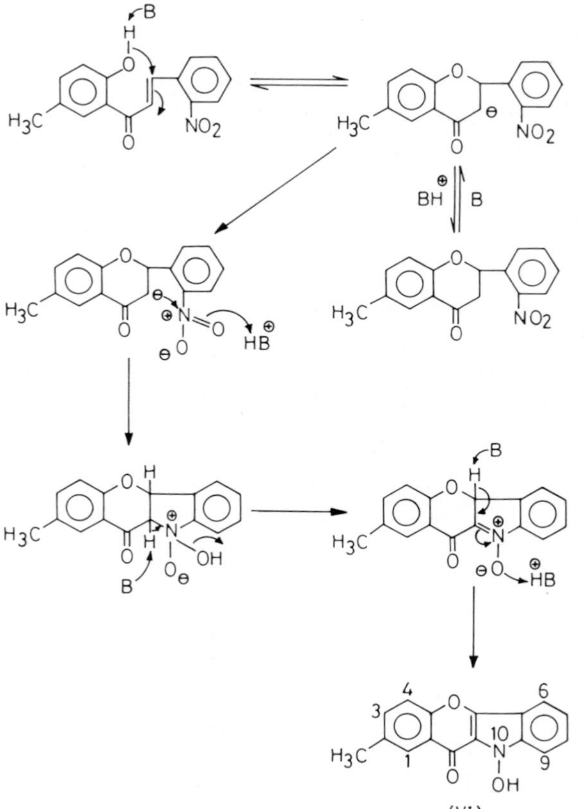

(VI)

The mechanism for the formation of **VI** has been suggested as follows[88]:

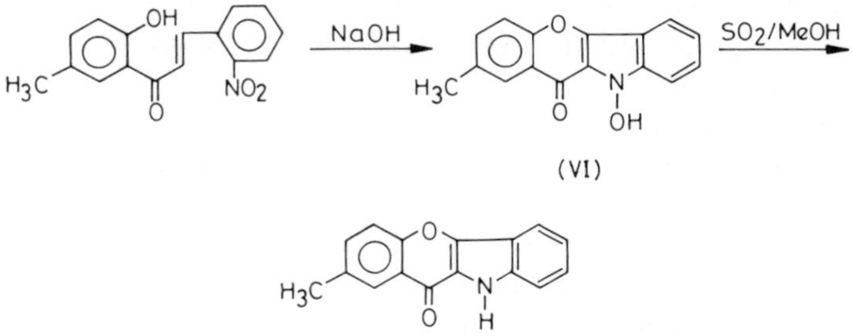

(VI)

(VII)

REFERENCES

1 Tatuta, H., *J. Chem. Soc. Japan*, **61**, 1048 (1940); *Chem. Abstr.*, **37**, 376^9 (1943).

2 Nakazawa, K., and Matsuura, S., *J. Pharm. Soc. Japan*, **75**, 469 (1955).

3 Matsuura, S., *Yakugaku Zasshi*, **77**, 296 (1957); *Chem. Abstr.*, **51**, 11337f (1957).

4 Akaboshi, S., and Kutsuma, T., *Yakugaku Zasshi*, **88**, 1020 (1968); *Chem. Abstr.*, **70**, 11526x (1969).

5 Vyas, G. N., and Shah, N. M., *J. Indian Chem. Soc.*, **26**, 273 (1945).

6 Vyas, G. N., and Shah, N. M., *Curr. Sci. (India)*, **19**, 318 (1950).

7 Seshadri, S., and Trivedi, P. L., *J. Org. Chem.*, **22**, 1633 (1957).

8 Tatsuta, H., *J. Chem. Soc. Japan*, **63**, 935 (1942).

9 Narasimhachari, N., and Seshadri, T. R., *Proc. Indian Acad. Sci.*, **29A**, 265 (1949).

10 Datta, S. C., Murti, V. V. S., and Seshadri, T. R., *Indian J. Chem.*, **9**, 614 (1971).

11 Jha, B. C., and Amin, G. C., *Tetrahedron*, **2**, 241 (1958).

12 Covello, M., Dini, A., and De Simone, F., *Rend. Accad. Sci. Fis. Mat., Naples*, **35**, 728 (1968); *Chem. Abstr.*, **74**, 99795a (1971).

13 Naik, H. B., Mankiwala, S. C., and Thakor, V. M., *J. Indian Chem. Soc.*, **52**, 1099 (1975).

14 Matsuura, S., Iinuma, M., Ito, T., Kuroiwa, M., and Higashi, N., *Yakugaku Zasshi*, **96**, 393 (1976); *Chem. Abstr.*, **85**, 5455s (1976).

15 Soni, V. T., Shah, R. V., and Amin, G. C., *Vidya*, **19B**, 205 (1976); *Chem. Abstr.*, **88**, 50598p (1978).

16 Yamato, M., and Hashigaki, K., Japan Kokai Tokkyo Koho 78121763 (1978); *Chem. Abstr.*, **90**, 87277a (1979).

17 Saiyad, I. Z., Nadkarni, D. R., and Wheeler, T. S., *J. Chem. Soc.*, 1737 (1937).

18 Anteunis, M., *Verhandel. Koninkl. Vlaam. Acad. Wetenschap., Belg., Kl. Wetenschap.*, **22**, No. 64, 1 (1960); *Chem. Abstr.*, **55**, 19917g (1961).

19 Raval, A. A., and Shah, N. M., *J. Org. Chem.*, **21**, 1408 (1956).

20 Dhar, D. N., and Lal, J. B., *J. Org. Chem.*, **23**, 1159 (1958).

21 Széll, T., *Chem. Ber.*, **91**, 2609 (1958).

22 Mulchandani, N. B., and Shah, N. M., *Chem. Ber.*, **93**, 1918 (1960).

23 Awad, W. I., El-Neweihy, M. F., and Selim., S. F., *J. Org. Chem.*, **25**, 1333 (1960).

24 Dhar, D. N., *J. Org. Chem.*, **25**, 1247 (1960).

25 Reichel, L., and Steudel, J., *Annalen*, **553**, 83 (1942).

26 Reichel, L., and Marchand, J., *Chem. Ber.*, **76B**, 1132 (1943).

27 Reichel, L., and Schickle, R., *Chem. Ber.*, **76B**, 1134 (1943).

28 Bognár, R., Tokes, A. L., and Frenzel, H., *Acta Chim.* (*Budapest*), **61**, 79 (1969); *Chem. Abstr.*, **71**, 113217k (1969).

29 Bognár, R., Farkas, I., and Rakoshi, M., *Acta Chim. Acad. Sci., Hun.*, **30**, 87 (1962); *Chem. Abstr.*, **58**, 4635c (1963).

30 Aurnhammer, G., Wagner, H., Hörhammer, L., and Farkas, L., *Chem. Ber.*, **104**, 473 (1971).

31 Mahal, H. S., Rai, H. S., and Venkataraman, K., *J. Chem. Soc.*, 866 (1935).

32 Balaiah, V., Row, L. R., and Seshadri, T. R., *Proc. Indian Acad. Sci.*, **20A**, 274 (1944).

33 Grouiller, A., and Thomassery, V., *Bull. Soc. Chim. France*, **12**, Pt. 2, 3448 (1973); *Chem. Abstr.*, **81**, 12959e (1974).

34 Lin, C-H., and Chen, F-C., *Formosan Sci.*, **12**, 5 (1958); *Chem. Abstr.*, **53**, 7159h (1959).

35 Panasenko, A. I., Kachurin, O. I., and Starkov, S. P., *Izv. Vyssh. Ucheb. Zaved., Khim. Khim. Tekhnol.*, **18**, 1203 (1975); *Chem. Abstr.*, **83**, 205454m (1975).

36 Panesenko, A. I., *Izv. Vyssh. Ucheb. Zaved., Khim. Khim. Tekhnol.*, **20**, 1786 (1977); *Chem. Abstr.*, **88**, 104325t (1978).

37 Seikel, M. K., Lounsbury, M. J., and Wang, S. C., *J. Org. Chem.*, **27**, 2952 (1962).

38 Welcomme, N., Lesieur, D., Lespagnol, C., Rabaron, A., and Navarro, J., *Ann. Pharm. Fr.*, **36**, 135 (1978); *Chem. Abstr.*, **89**, 190991g (1978).

39 Ferreira, D., Bradt, E. V., Volstedt, F. du R., and Roux, D. G., *J. Chem. Soc. Perkin Trans.*, **1**, 1437 (1975).

40 Naik, V. G., and Marathey, M. G., *J. Univ. Poona, Science Technol.*, **18**, 61 (1960); *Chem. Abstr.*, **55**, 1599a (1961).

41 Gore, K. G., Naik, V. G., and Marathey, M. G., *J. Univ. Poona, Sci. Technol.*, **18**, 65 (1960); *Chem. Abstr.*, **55**, 1599e (1961).

42 Gore, K. G., Naik, V. G., and Marathey, M. G., *J. Univ. Poona, Sci. Technol.*, **18**, 69 (1960); *Chem. Abstr.*, **55**, 1599h (1961).

43 Raut, K. B., *Dissertation Abstr.*, **20**, 97 (1959).

44 Raut, K. B., and Wender, S. H., *J. Org. Chem.*, **25**, 50 (1960).

45 Cullen, W. P., Donnelly, D. M. X., Keenan, A. K., Keenan, P. J., and Ramdas, K., *J. Chem. Soc., Perkin Trans., I*, 1671 (1975).

46 Marathey, M. G., *J. Univ. Poona, Sci. Technol.*, **18**, 53 (1960); *Chem. Abstr.*, **55**, 1598e (1961).

47 Naik, V. G., and Marathey, M. G., *J. Univ. Poona, Sci. Technol.*, **18**, 55 (1960); *Chem. Abstr.*, **55**, 1598e (1961).

48 Gore, K. G., Sonaware, H. R., Thankar, G. P., and Marathey, M. G., *J. Univ. Poona, Sci. Technol.*, **18**, 73 (1960); *Chem. Abstr.*, **55**, 1600c (1961).

49 Jacob, K. C., Jadhav, G. V., and Vakharia, M. N., *Pesticides*, **7**, 23 (1973).

50 Vernekar, S. S., and Rajagopal, S., *Rec. Trav. Chim.*, **81**, 710 (1962); *Chem. Abstr.*, **58**, 5559g (1968).

51 Ambekar, S. Y., Dandeganoker, S. H., Jolad, S. D., and Rajagopal, S., *J. Indian Chem. Soc.*, **40**, 1041 (1963).

52 Algar, J., and Flynn, J. P., *Proc. Roy. Irish Acad.*, **42B**, 1 (1934).

53 Oyamada, T., *J. Chem. Soc. Japan*, **55**, 1256 (1934); *Bull. Chem. Soc. Japan*, **10**, 182 (1935).

54 Narasimhachari, N., Rajagopalan, D., and Seshadri, T. R., *Proc. Indian Acad. Sci.*, **37A**, 705 (1953).

55 Geissman, T. A., and Fukushina, D. K., *J. Am. Chem. Soc.*, **70**, 1686 (1948).

56 Narasimhachari, N., and Seshadri, T. R., *Proc. Indian Acad. Sci.*, **30A**, 216 (1949).

57 Cullen, W. P., Donnelly, D. M. X., Keenan, A. K., Lavin, T. P., Melody, D. P., and Philbin, E. M., *J. Chem. Soc., C*, 2848 (1971).

58 Simpson, T. H., and Whalley, W. B., *J. Chem. Soc.*, 166 (1955).

59 Anand, N., Iyer, R. N., and Venkataraman, K., *Proc. Indian Acad. Sci.*, **29A**, 203 (1949).

60 O'Sullivan, W. I., Brady, B. A., and Philbin, E. M., *J. Chem. Soc., D*, 1435 (1970).

61 Bellino, A., and Venturella, P., *Atti. Accad. Sci., Lettere Arti Palermo*, **21**, 17 (1962); *Chem. Abstr.*, **59**, 552d (1963).

62 Seshadri, S., and Trivedi, P. L., *J. Org. Chem.*, **25**, 841 (1960).

63 Cummins, B., Donnelly, D. M. X., Philbin, E. M., Swirski, J., Wheeler, T. S., and Wilson, R. K., *Chem. Ind. (London)*, 348 (1960).

64 Donnelly, D. M. X., Eades, J. F., Philbin, E. M., and Wheeler, T. S., *Chem. Ind. (London)*, 1453 (1961).

65 Cummins, B., Donnelly, D. M. X., Eades, J. F., Fletcher, H., O'Cinneide, F., Philbin, E. M., Swirski, J., Wheeler, T. S., and Wilson, R. K., *Tetrahedron*, **19**, 499 (1963).

66 Gromley, T. R., and O'Sullivan, W. I., *Tetrahedron*, **29**, 369 (1973).

67 Budagyants, M. I., Shakhova, M. K., and Samokhvalov, G. I., *Zh. Organ. Khim.*, **5**, 1857 (1969); *Chem. Abstr.*, **72**, 21600x (1970).

68 Kelkar, A. S., and Kulkarni, A. B., *Indian J. Chem.*, **11**, 726 (1973).

69 Jerzmanowska, Z., and Podwinski, B., *Roczniki Chem.*, **42**, 307 (1968).

70 Nadkarni, D. R., and Wheeler, T. S., *J. Chem. Soc.*, 1320 (1938).

71 Matsuura, S., and Kunii, T., *Yakugaku Zasshi*, **94**, 645 (1974); *Chem. Abstr.*, **81**, 63440m (1974).

72 Hsu, K. K., Shi, J. Y., *J. Chem. Soc., Taipei*, **20**, 51 (1973); (*Chem. Abstr.*, **78**, 159370c (1973).

73 Young, A. N., *Diss. Abstr., Int.*, **33B**, 2536 (1972).

74 Matsuura, S., Kunii, T., and Matsuura, A., *Chem. Pharm. Bull.*, **21**, 2757 (1973); *Chem. Abstr.*, **80**, 59824b (1974).

75 Chem, F-C., Veng, T., Chen, C. Y., and Chang, P. W., *J. Chin. Chem. Soc. (Taipei)*, **17**, 251 (1970); *Chem. Abstr.*, **74**, 87765c (1971).

76 Mahal, H. S., and Venkataraman, K., *J. Chem. Soc.*, 569 (1936).

77 Chakravarti, D., and Dutta, J., *J. Indian Chem. Soc.*, **16**, 639 (1939).

78 Joshi, K. C., and Jauhar, A. K., *Indian J. Chem.*, **1**, 477 (1963).

79 Grishko, L. G., Grabovskaya, V. V., Marchuk, L. A., and Khilya, V. P., *Dopov. Akad. Nauk Ukr. RSR, Ser. B, Geol., Khim. Biol. Nauki*, (5), 426 (1978); *Chem. Abstr.*, **89**, 108956t (1978).

80 Jain, A. C., and Khazanchi, R., *Indian J. Chem.*, **16B**, 1125 (1978).

81 Ahmad, S., Wagner, H., and Razaq, S., *Tetrahedron*, **34**, 1593 (1978).

82 Massicot, J., *Compt. Rend.*, **240**, 94 (1955).

83 Bose, P. K., Chakrabarti, P., and Sanyal, A. K., *J. Indian Chem. Soc.*, **48**, 1163 (1971).

84 Kasahare, A., Izumi, T., and Ooshima, M., *Bull. Chem. Soc. Japan*, **47**, 2526 (1974).

85 Jain, A. C., Lal, P., and Seshadri, T. R., *Indian J. Chem.*, **7**, 305 (1969).

86 Farkas, L., and Wolfner, A., *Magy. Kem. Foly.*, **81**, 357 (1975); *Chem. Abstr.*, **83**, 178725s (1975).

87 Farkas, L., and Wolfner, A., *Flavonoids Bioflavonoids, Proc. Hung. Bioflavonoids Symp., 5th*, pp. 193–6 (1977). Edited by L. Farkas, M. Gabor, and F. Kallay, Elsevier, Amsterdam, Netherlands.

88 Dean, F. M., Patampongse, C., and Podimuang, V., *J. Chem. Soc. Perkin Trans.*, **1**, 583 (1974).

89 Wurm, G., *Arch. Pharm. (Weinheim)*, **306**, 299 (1973); *Chem. Abstr.*, **79**, 66126h (1973).

90 Bhardwaj, D. K., Jain, S. C., and Sharma, G. C., *Indian J. Chem.*, **15B**, 860 (1977).

91 Weber, F. G., and Westphal, G., *Pharmazie*, **30**, 283 (1975); *Chem. Abstr.*, **83**, 96951y (1975).

Chapter Fourteen
Biochemical Cyclization and Degradation of Chalcones

FORMATION OF FLAVANONES

The interconversion of chalcone and flavanone with a phloroglucinol type of substitution has been accomplished at pH 7 by the aid of an enzyme isolated from the peel of *Citrus aurantium.*[1] Other sources of the enzyme are the peels of *C. natsudaidai, C. junos, C. nobilis, C. pseudoparadisi,* and *Poncirus trifoliata.*

    2′,4,4′-Trihydroxychalcone has been successfully isomerized[2] to the optically active 4′,7-dihydroxyflavanone by the mediation of the isomerase isolated from soyabean seedling[2]:

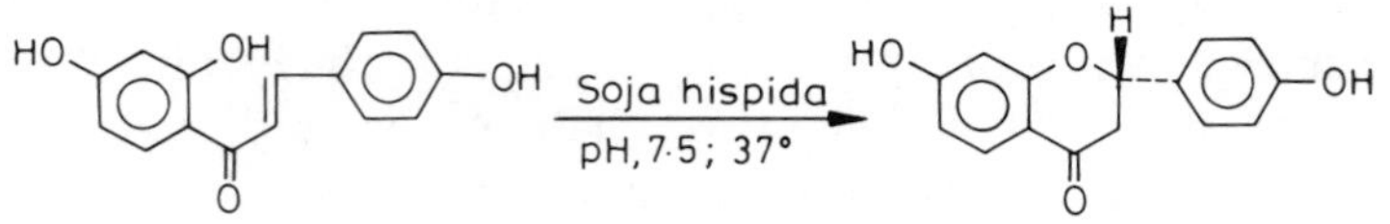

However, the chalcone–flavanone isomerase isolated either from the young leaves of *Datisca cannabina*[3] or from three different seedlings,[4] *Phaseolus aureus, Cicer arietinum,* and *Petroselinum hortense,* has failed in its action on chalcone glucoside, including those having a resorcinol type[3] of substitution in ring A. It is interesting that the chalcone isomerase isolated either from the flowers of Apeldoorn tulips or *Lilium candidum*[5] is active against 2′,4,4′,6′-tetrahydroxychalcone but is ineffective against 2′,4,4′-trihydroxychalcone.

Several studies have been reported about the chalcone–flavanone isomerase activity in plants.[6-9] The chalcone–flavanone isomerase obtained from soyabean seeds has been purified.[10] The kinetics of the isomerization reaction—isoliquiritigenin to liquiritigenin—conducted under the influence of the aforesaid enzyme has been examined.[10]

FORMATION OF BENZALCOUMARANONE

Enzyme-catalyzed transformation, at pH 5–6, of hydroxychalcone glycoside to hydroxybenzalcoumaranone glycoside has been reported.[11] The required enzyme is obtained by alcohol extraction of the macerated rays of *Coreopsis lanceolata.*

FORMATION OF FLAVONOL AND AURONE

Under aerobic conditions and in the presence of trace quantity of hydrogen peroxide isoliquiritigenin (2′,4,4′-trihydroxychalcone) undergoes catalyzed oxidation by horseradish peroxidase[12] to yield 4′,7-dihydroxyflavonol and 4′,6-dihydroxyaurone. A similar reaction occurs under the influence of cell-free extracts of hypocotyl obtained from *Phaseolus vulgaris.*

FORMATION OF ANTHOCYANIN AND OTHER FLAVONOIDS

It has been established that 2′,3,4,4′,6′-pentahydroxychalcone-4′-glucoside serves as a precursor in the synthesis of anthocyanin[13,14] and other flavonoids[16,17]. 2′,4,4′-Trihydroxychalcone is reported to be a precursor in the biosynthesis of coumestrol in the seedlings of *Phaseolus aureus.*[18] Further, it has been shown that there is maximum incorporation of the chalcone into amorphigenin by germinating *Amorpha frutucosa* seeds.[19]

DEGRADATION OF CHALCONES

A few cases of degradation by chalcones brought about by plant cell suspension cultures have been reported.[20,21] For example, 2',4,4',6'-tetrahydroxychalcone-2'-β-D-glucoside[21] undergoes degradation by callus suspension cultures of *Pisum sativum* L. into *para*-hydroxybenzoic acid and 3-(hydroxyphenyl)prop-2-enoic acid.

The chalcone content in the peel of pineapple fruit is reported to decrease with maturity.[22]

REFERENCES

1   Shimokoriyama, M., *J. Am. Chem. Soc.*, **79**, 4199 (1957).

2   Wong, E., and Moustafa, E., *Tetrahedron Letters*, 3021 (1966).

3   Grambow, H. J., and Grisebach, H., *Phytochemistry*, **10**, 789 (1971).

4   Hahlbrock, K., Wong, E., Schill, L., and Grisebach, H., *Phytochemistry*, **9**, 949 (1970).

5   Wiermann, R., *Planta*, **102**, 55 (1972).

6   Forkmann, G., and Kuhn, B., *Planta*, **144**, 189 (1979).

7   Boland, M. J., and Wong, E., *Bioorg. Chem.*, **8**, 1 (1979).

8   Kuhn, B., Forkmann, G., and Seyffert, W., *Planta*, **138**, 199 (1978).

9   Weissenboeck, G., and Sachs, G., *Planta*, **137**, 49 (1977).

10 Boland, M. J., and Wong, E., *Eur. J. Biol.*, **50**, 383 (1975).

11 Shimokoriyama, M., and Hattori, S., *J. Am. Chem. Soc.*, **75**, 2277 (1953).

12 Rathmal, W. G., and Bendal, D. S., *Biochem. J.*, **127**, 125 (1972).

13 Endress, R., *Phytochemistry*, **13**, 421 (1974).

14 Endress, R., *Phytochemistry*, **13**, 599 (1974).

15 Endress, R., *Z. Pflanzenphysiol.*, **67**, 188 (1972); *Chem. Abstr.*, **77**, 58895b (1972).

16 Wong, E., *Phytochemistry*, **7**, 1751 (1968).

17 Wong, E., and Grisebach, H., *Phytochemistry*, **8**, 1419 (1969).

18 Dewick, P. M., Barz, W., and Grisebach, H., *Phytochemistry*, **9**, 775 (1970).

19 Crombie, L., Dewick, P. M., and Whiting, D. A., *J. Chem. Soc.*, D, 1183 (1971).

20 Berlin, J., Kiss, P., Mueller-Enoch, D., Gierse, D., Barz, W. and Janistyn, B., *Z. Naturforsch.*, **29C**, 374 (1974).

21 Janistyn, B., Barz, W., and Pohl, R., *Z. Naturforsch.*, **26B**, 973 (1971).

22 Lodh, S. B., Divakar, N. G., Chadha, K. L., Melanta, K. R., and Selvaraj, Y., *Indian J. Hort.*, **30**, 381 (1973).

Chapter Fifteen

Miscellaneous Reactions of Chalcones

CAUSTIC ALKALI[1]

The olefinic bond in the chalcone molecule undergoes cleavage when refluxed with concentrated alkali (i.e., 0.3 M)[2] giving rise to acetophenone and benzaldehyde. A kinetic study[3] of the above reaction has been carried out, and a mechanism[2,4] has been put forward to explain the formation of these products:

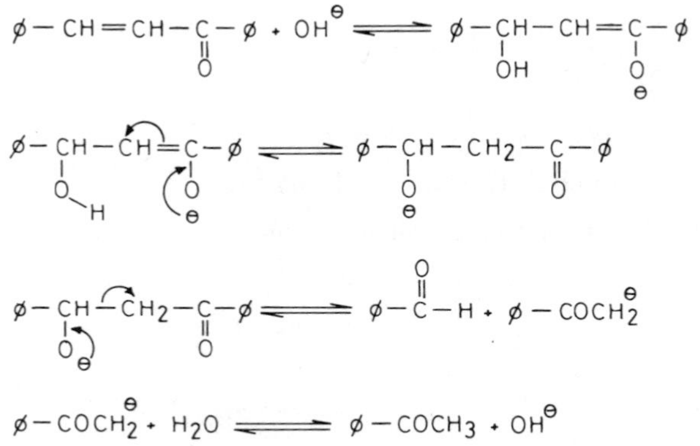

The rate of the reaction is proportional to the concentration of base,[3] and the rate-determining step is the attack by the hydroxide ions.[3,5]

In the case of some chalcones higher molecular weight compounds have been isolated.[6] Thus anisal bis(3,4-dimethoxyacetophenone)[6] is produced by the reaction of caustic alkali with 3',4,4'-trimethoxychalcone.

SULFURIC ACID

Chalcones usually show halochromic effects when wetted with concentrated sulfuric acid (see Halochromism, Chapter 18). Some chalcones remain unchanged on this treatment, while others are susceptible to sulfonation in ring B.[7] Thus, 4-methoxychalcone and 4,4'-dimethoxychalcone are converted in the above reaction to their corresponding sulfonic acid derivatives, while 4'-methoxychalcone remains unaffected.

The following reactions take place when chalcone is reacted with sulfuric acid.[7] The first step is the protonation of the carbonyl oxygen, followed by the slow formation of monosulfonic acid[8]:

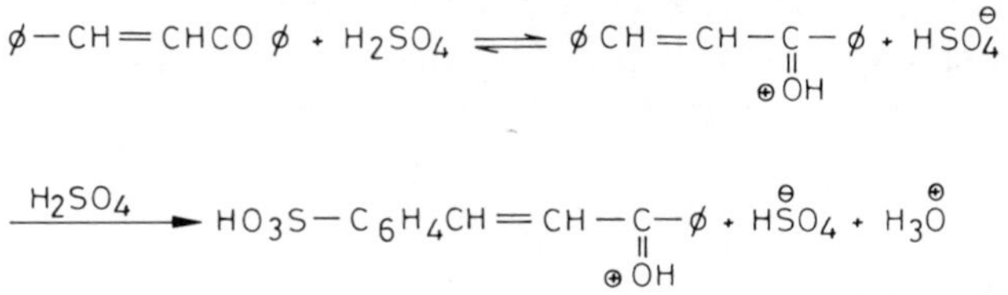

$$\phi-CH=CHCO\,\phi \; + \; H_2SO_4 \; \rightleftharpoons \; \phi\,CH=CH-\underset{\underset{\oplus OH}{\|}}{C}-\phi \; + \; HSO_4^{\ominus}$$

$$\xrightarrow{H_2SO_4} \; HO_3S-C_6H_4CH=CH-\underset{\underset{\oplus OH}{\|}}{C}-\phi \; + \; H\overset{\ominus}{S}O_4 \; + \; H_3\overset{\oplus}{O}$$

POLYPHOSPHORIC ACID (PPA)

Chalcone is transformed into 3-phenylindan-1-one (50%) when it is heated with polyphosphoric acid[9]:

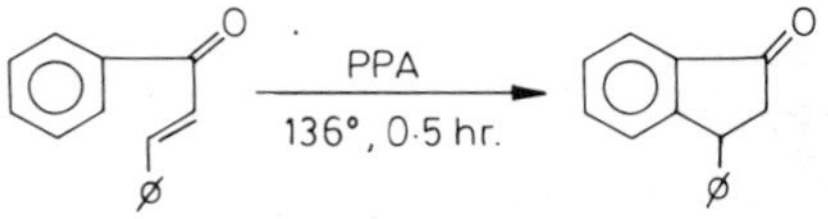

Two side reactions, one leading to the formation of dihydrochalcone and the other to an aromatic acid (formed by α-carbonyl cleavage), have been observed.[9]

PHENYLACETIC ACID

Chalcone reacts with disodiophenylacetic acid (prepared by the interaction of the acid with 2 equiv of sodium amide in liquid ammonia) to give the corresponding keto acid[10] in excellent yield:

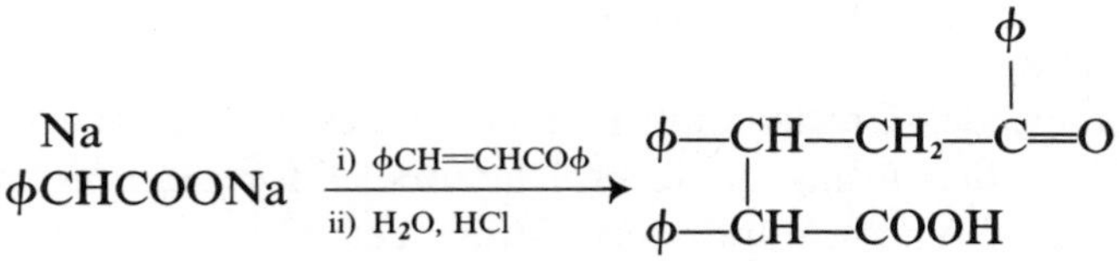

$$\begin{array}{c} \text{Na} \\ \phi\text{CHCOONa} \end{array} \xrightarrow[\text{ii) H}_2\text{O, HCl}]{\text{i) } \phi\text{CH=CHCO}\phi} \begin{array}{c} \overset{\displaystyle \phi}{|} \\ \phi-\text{CH}-\text{CH}_2-\text{C}=\text{O} \\ | \\ \phi-\text{CH}-\text{COOH} \end{array}$$

SODIUM BISULFITE

2'-Hydroxychalcone is reported to react (at 135°) with bisulfite to yield 21% of 2-(o-hydroxybenzoyl)-1-phenylethane sulfonic acid.[11] 4'-Hydroxy-2'-methoxychalcone and 4'-hydroxy-3,3',4-trimethoxychalcone react with sodium bisulfite in an analogous fashion.

ALKALI SULFIDE

Chalcone hydrosulfide,[12] ϕCH=CH—C(OH)SH, is produced when chalcone, in alcoholic potassium hydroxide, is saturated with hydrogen sulfide. Dibenzalacetophenone disulfide,[12] however, results when chalcone and NaSH are treated in an alcoholic solution. trans-α-Bromomethylchalcone reacts with sodium hydrogen sulfide to yield the following products[13]:

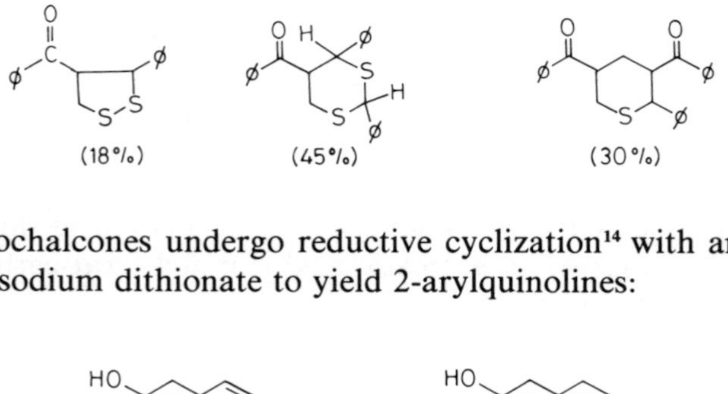

6-Nitrochalcones undergo reductive cyclization[14] with ammonium sulfide or sodium dithionate to yield 2-arylquinolines:

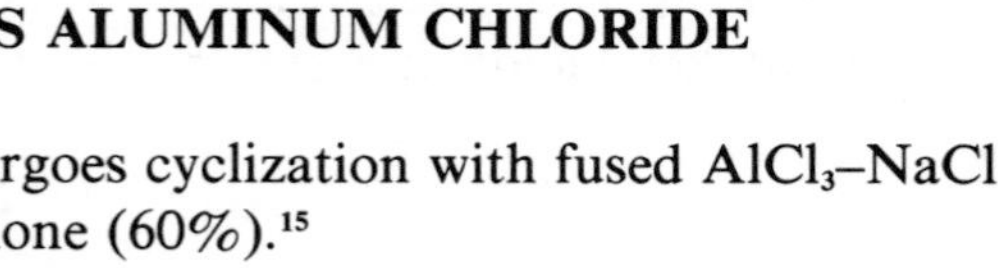

ANHYDROUS ALUMINUM CHLORIDE

Chalcone undergoes cyclization with fused $AlCl_3$–NaCl to yield the 3-phenyl-1-indanone (60%).[15]

Under Friedel–Crafts conditions chalcones in general yield different products, depending on several factors,[16,17] for example, time and temperature of reaction and steric influences. Chalcone on treatment with anhydrous $AlCl_3$ in benzene yields $\beta\beta'$-diphenylpropiophenone (90%)[17] together with small amounts of 3-phenylhydrindone[17] and *cis*-chalcone.[18] On the other hand, α-methyl- and α-phenylchalcones under similar conditions yield the corresponding hydrindones. In these two cases apparently the steric effects[17] are responsible for the formation of hydrindone in preference to the addition product. There are also reports about the preparation of $\beta\beta'$-disubstituted propiophenones[19] and/or of substituted hydrindones[20] based on the above reaction. Aluminum chloride is also reported to bring about partial demethylation.[21] Thus 2'-hydroxy-4',6'-dimethoxychalcone, under the above conditions, yields 2',4'-dihydroxy-6'-methoxychalcone.

Furthermore, chalcone and chlorobenzene react in the presence of $AlCl_3$ to give products as indicated[22]:

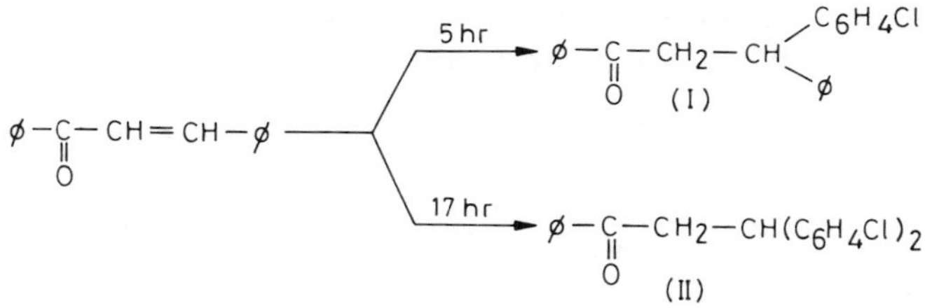

In the formation of **I** the usual addition occurs across the C=C of the chalcone molecule, and the aromatic residue, C_6H_4Cl, is attracted to the β-carbon atom. When the reaction period is prolonged, then the replacement of the phenyl group by chlorophenyl residue takes place at the β-carbon atom, leading to the formation of **II**.

NITROSOYL CHLORIDE[23]

At room temperature chalcone adds to a molecule of nitrosoyl chloride, and a good yield of the adduct is obtained.

SODIUM AMIDE[24]

2',4,4'-Trimethoxychalcone on treatment with sodium amide in refluxing toluene undergoes molecular cleavage to yield 1-amido-2-(*p*-methoxyphenyl)ethylene and 1,3-dimethoxybenzene.

MERCURY SALTS

Mercury chloride and mercury bromide are reported to yield addition products with chalcone.[25] α-Mercuration occurs when chalcone reacts with mercuric acetate in the presence of t-butyl hydroperoxide and perchloric acid.[26]

$$\phi CH{=}CHCO\phi \xrightarrow{\text{t-butylperoxymercuration}} \phi CH(O{-}OBu^t)CH(HgOAc)CO\phi$$
$$(77\%)$$

A parallel pattern of behavior is exhibited by chalcone in methoxymercuration reaction.[26] The kinetics of methoxymercuration of substituted chalcones has been studied.[27] The following mechanism has been postulated[27] for the above-mentioned reaction:

$$(Bz = C_6H_5CO-)$$

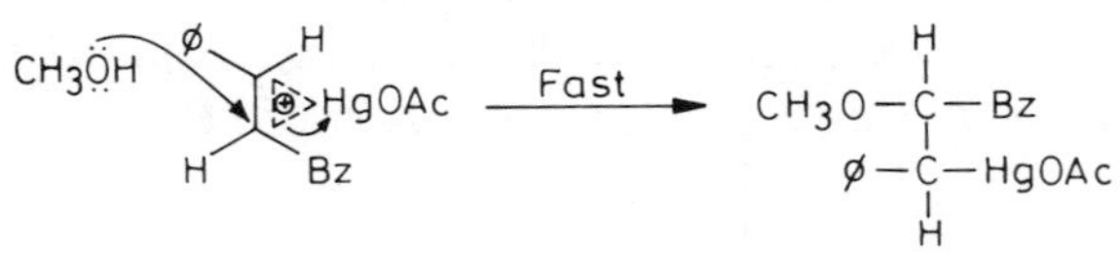

IRON CARBONYL

The preparation of several iron carbonyl (tri- and tetra-) complexes of ferrocene analogues of chalcone has been described.[28] The introduction of the iron carbonyl moiety is reported to interfere with the conjugation in the molecule of ferrocene analogue of chalcone.

Iron carbonyl, $Fe_2(CO)_9$, reacts with chalcone to yield a complex.

The iron carbonyl–chalcone complex reacts with triphenylphosphine[29] in accordance with the reaction:

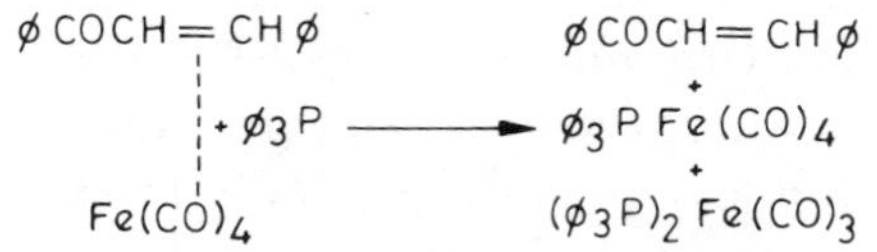

$$\phi COCH = CH\phi \atop \underset{Fe(CO)_4}{|} + \phi_3 P \longrightarrow \phi COCH = CH\phi \atop \phi_3 P \overset{+}{Fe}(CO)_4 \atop (\phi_3 P)_2 Fe(CO)_3$$

TRITIUM[30]

Hydrogenation of 2'-D-glucosyl-4,4',6'-trihydroxychalcone with tritium yields **III**:

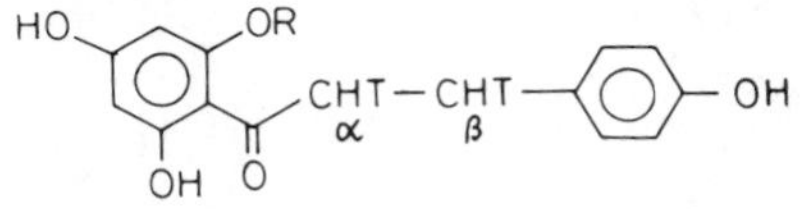

(III)

The incorporation of tritium in the dihydro compound (**III**) is reported to be of the order of 60% at the α-carbon atom (with respect to the carbonyl group) and 15–20% at the β-carbon atom.[30]

HEAVY WATER[31]

The α-hydrogen of chalcone does not exchange for deuterium when it is heated with heavy water–dioxane in the presence of a basic catalyst.

DEUTERIOETHANOL

Chalcone is reported to undergo base-catalyzed deuteration with deuterioethanol to yield the α-deuterioketone (42%).[32]

BENZENE AND NITROPARAFFINS

α-Substituted chalcone derivatives are reported to undergo addition reaction with benzene in the presence of palladium (**II**) acetate.[33]

Chalcones react with nitroparaffins,[38,39] for example, nitromethane,[34-37] 1- (and 2-) nitropropanes[35] and *gem*-dinitroalkanes,[40] to give Michael-type addition compounds, generally in excellent yields.[38] This reaction has been accomplished under the influence of basic catalysts, namely, alcoholic ammonia,[34] calcium hydride in methanol,[35] or pyridine.[40]

The primary addition product of chalcone and nitromethane,[36] for example, is 4-nitro-1,3-diphenylbutan-1-one (**IV**), further reaction yielding the bis adduct (**V**):

$$PhCH\text{---}CH_2COPh \qquad\qquad Ph\text{---}CH\text{---}CH_2COPh$$
$$| \qquad\qquad\qquad\qquad\qquad\qquad\qquad |$$
$$CH_2\text{---}NO_2 \qquad\qquad\qquad\qquad CHNO_2$$
$$\qquad\qquad\qquad\qquad\qquad\qquad\qquad |$$
$$\qquad\qquad\qquad\qquad\qquad Ph\text{---}CH\text{---}CH_2COPh$$
$$\textbf{(IV)} \qquad\qquad\qquad\qquad\qquad\qquad \textbf{(V)}$$

DIAZOMETHANE

Chalcone reacts additively with diazomethane to yield two isomeric pyrazolines,[41] which upon heating pass into cyclopropane derivative.[42] Homologation of chalcone, however, occurs[43] without significant pyrazoline formation if the reaction with diazomethane is performed in the presence of fluoroboric acid.

1,1-DIETHYLAMINOPROP-1-YNE

Chalcone is reported to react with 1,1-diethylaminoprop-1-yne to yield a pyran derivative (**VI**), which on hydrolysis gives the lactone **VII** and keto acid **VIII**.[44]

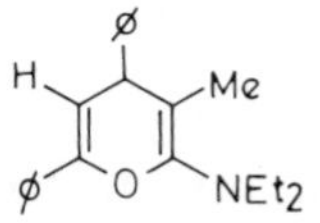

       (VI) 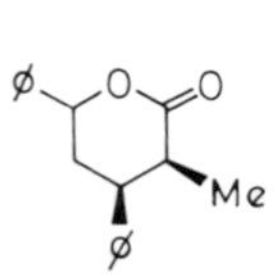   (VII) 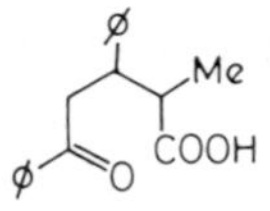   (VIII)

ALCOHOLS

1 Chalcones react with nitroethyl- and nitroisopropyl alcohols in liquid ammonia according to the scheme[45]:

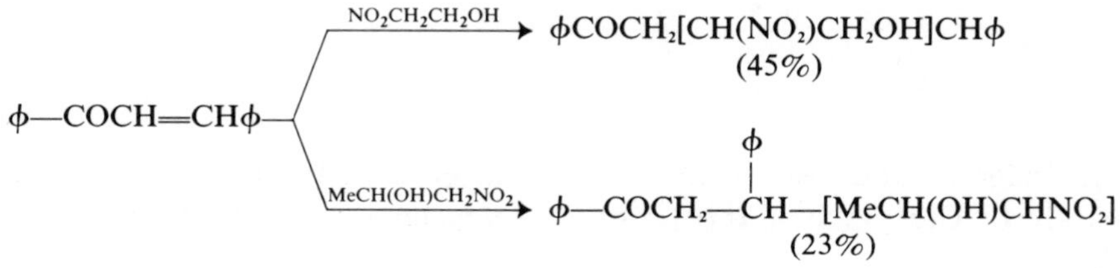

$$\phi\text{COCH}_2[\text{CH(NO}_2)\text{CH}_2\text{OH}]\text{CH}\phi$$
$$(45\%)$$

$$\phi\text{—COCH}=\text{CH}\phi$$

via $\text{NO}_2\text{CH}_2\text{CH}_2\text{OH}$

via $\text{MeCH(OH)CH}_2\text{NO}_2$

$$\phi\text{—COCH}_2\text{—}\overset{\displaystyle\phi}{\overset{|}{\text{CH}}}\text{—[MeCH(OH)CHNO}_2]$$
$$(23\%)$$

2 The selective hydrogenation of chalcone has been achieved by using ethylene glycol in the presence of a catalyst, $\text{RuCl}_2\cdot\text{P}\phi_3$.[46] The yield of α,β-saturated ketones[46] ranges between 77 and 99%. As an extension of this reaction, chalcone has been used as a solvent–hydrogen acceptor[47] in the presence of $\text{RuCl}_2\cdot\text{P}\phi_3$ for effecting dehydrogenation of α-ethylenic alcohols and glycols.

THIOLS[48,49]

Excellent yields of keto sulfides (**IX**) are reported to be obtained when substituted chalcone and the hydrochlorides of 2-diethylaminoethyl- and 3-diethylaminopropylmercaptans are allowed to react[48]:

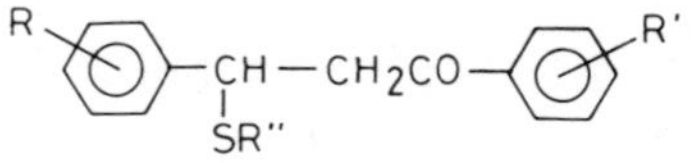

$$(\text{IX}) \qquad R''= \text{Et}_2\text{NCH}_2\text{CH}_2\text{—};$$
$$\text{Et}_2\text{NCH}_2\text{CH}_2\text{CH}_2\text{—}$$

Pentanethiol, benzenethiol, and toluenethiol likewise react with ferrocene analogue of chalcone to give their corresponding adducts.[50]

2-Aminoethanethiol,[51] however, reacts with chalcone to yield either a monoadduct, **X** and **XI**, or the bis compound, depending on the molar proportion of the reactants used:

$$\phi COCH_2 \cdot CH\phi \cdot S \cdot (CH_2)_2 \cdot NH_2 \qquad \textbf{(X)}$$
$$\phi COCH_2 \cdot CH\phi \cdot NH \cdot (CH_2)_2 \cdot SH \qquad \textbf{(XI)}$$
$$\phi COCH_2 \cdot CH\phi \cdot S \cdot (CH_2)_2 \cdot NH \cdot CH\phi \cdot CH_2 CO\phi$$

SELENOLS

Arylselenols add readily to chalcone in ethanol in the absence of a catalyst. The yield of the product, ketoselenides,[52] vary from 44 to 80%.

3,5-DIMETHOXYPHENOL[53]

Substituted chalcone is reported to undergo β-coupling with 3,5-dimethoxyphenol in the presence of alkaline hydrogen peroxide, to yield the two structural isomers of 3,3-diaryl-2-hydroxypropiophenone[53]:

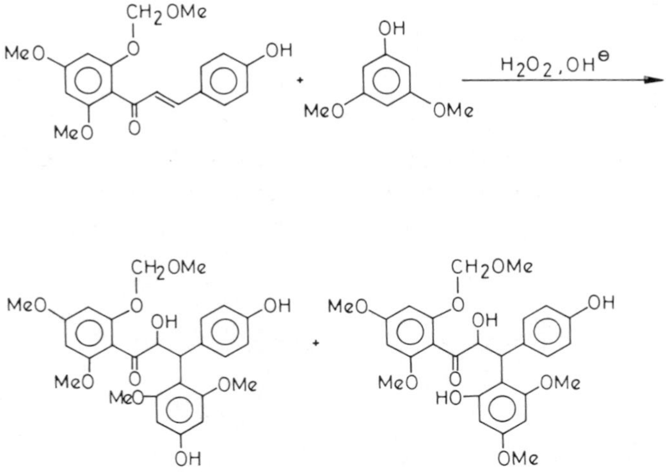

2-AMINOTHIOPHENOL

2,3-Dihydro-2,4-diphenyl-1,5-benzothiazepines[54] have been prepared by the reaction of 2-aminothiophenol with substituted chalcones.

KETENES

δ-Lactones[55] are obtained from chalcones by their interaction with diphenylketene quinoline, followed by an oxidative step. The preparation of α,α-diphenyl-β-methoxyphenyl-γ-benzoylbutyric lactone (XII) is illustrated[55]:

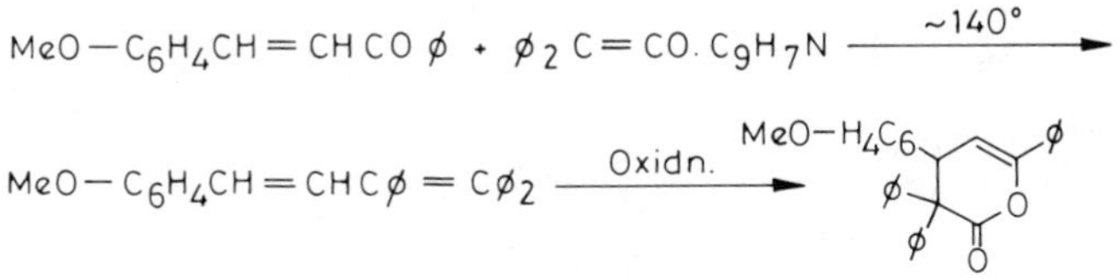

$$MeO-C_6H_4CH=CHCO\phi + \phi_2C=CO.C_9H_7N \xrightarrow{\sim 140°}$$

$$MeO-C_6H_4CH=CHC\phi=C\phi_2 \xrightarrow{Oxidn.}$$

(XII)

Chalcone is reported to react with 1 equiv of ketene acetal to yield 1,1-diethyl-2-benzoyl-3-phenylcyclobutane (XIII), which may be hydrolyzed to 3-β-phenyl-γ-benzoylbutyric acid (XIV).[56]

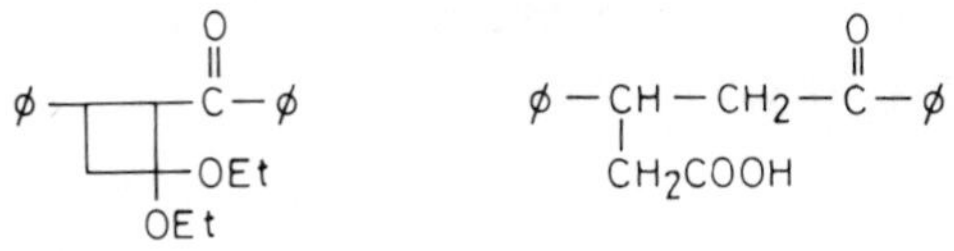

(XIII)   (XIV)

The Michael-type reaction of chalcone with o-silylated ketene furnishes the corresponding δ-keto esters[57] in good yields. The following is an illustrative example:

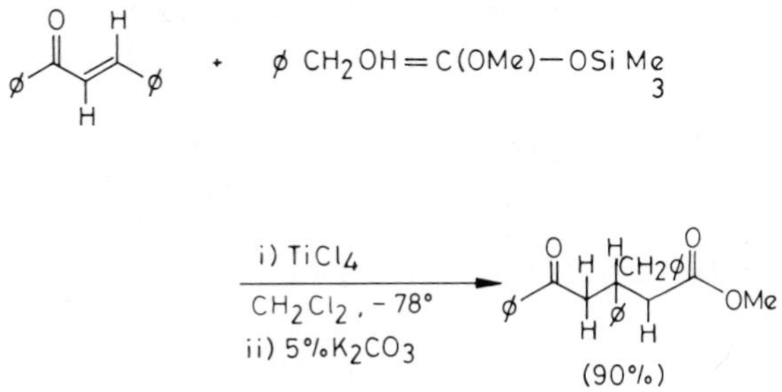

$$+ \quad \phi CH_2OH=C(OMe)-OSiMe_3$$

(90%)

BENZOYL CHLORIDE

Pyrylium salts[58] are formed when β-methylchalcones, in acetic acid, are reacted with benzoyl chloride in the presence of a Lewis acid. One of the possible mechanisms suggested[58] for the above reaction is outlined here[58]:

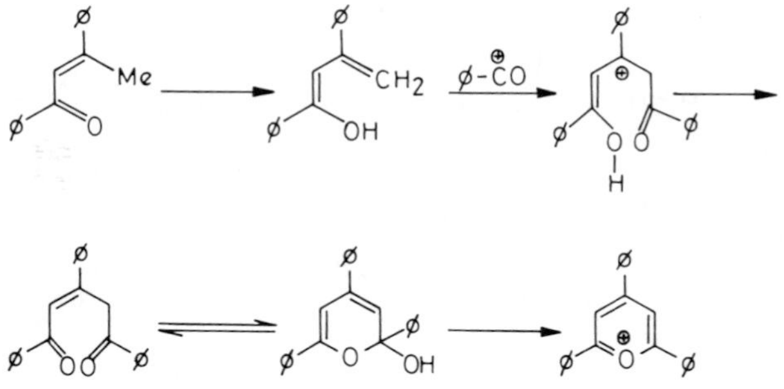

BENZYL-p-TOLYLSULFONE

In the presence of sodium ethoxide, chalcone is reported to condense with benzyl-p-tolylsulfone to yield a product[59] (XV, ~15%), which exists in two isomeric forms:

$$\phi-CH-CH_2-CO\phi$$
$$\overset{|}{\phi-CH-SO_2C_6H_4CH_3-p}$$
$$(\mathbf{XV})$$

SULFENYL COMPOUNDS

Chalcones react with phenyl sulfenyl chloride[60] in acetic acid medium to furnish an adduct (XVI) or thioaurones (XVII):

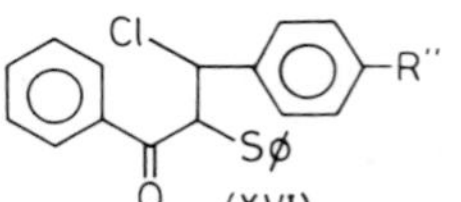

(XVI)

R″= H or NO₂

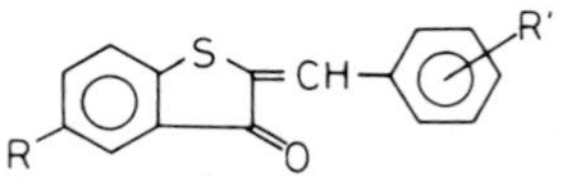

(XVII)

R = Me, Cl
R′= o−OH; o−Cl; o−MeO;
p−Me₂N and p−NH₂SO₂

1-(Phenylsulfinyl)-2,4-diphenyl-3-buten-2-ol[61] is preparable by the interaction of chalcone with $\phi S(O)CH_3$ in the presence of sodium amide in liquid ammonia.

DIMETHYLSULFONIUM AND SULFOXONIUM METHYLIDES

Dimethylsulfonium methylide is reported to react with chalcone to yield the corresponding oxirane by the selective addition of methylene to the carbonyl group[62]:

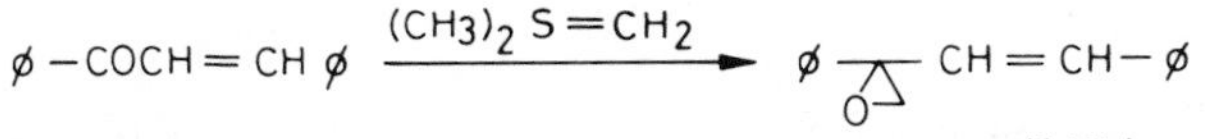

$$\phi-COCH=CH\,\phi \xrightarrow{(CH_3)_2\,S=CH_2} \phi \overset{}{\underset{O}{\triangle}} CH=CH-\phi$$

$$(84\%)$$

On the other hand, *trans*-chalcone reacts with dimethylsulfoxium methylide (DMSOM) leading to the formation of *trans*-1-phenyl-2-benzoylcyclopropane.[63] 2-Methoxychalcone, on prolonged treatment with DMSOM, however,yields **XVIII**, which cyclizes to pyran derivative[64] (**XIX**) in the presence of acid. Thus

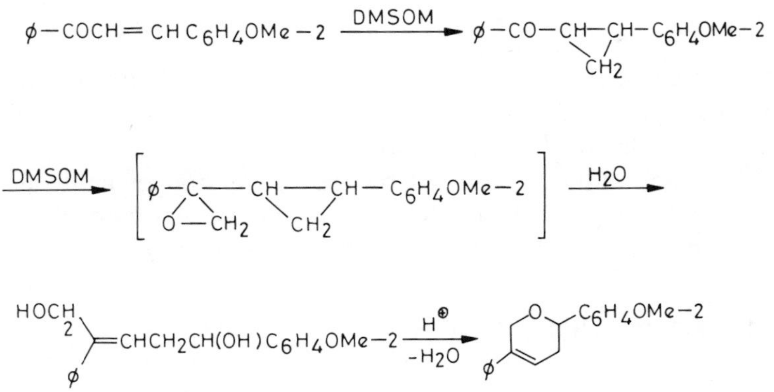

$$\phi-COCH=CH\,C_6H_4OMe-2 \xrightarrow{DMSOM} \phi-CO-\underset{\underset{CH_2}{\diagdown\diagup}}{CH-CH}-C_6H_4OMe-2$$

$$\xrightarrow{DMSOM} \left[\phi-\underset{O-CH_2}{\overset{}{C}}-CH-\underset{CH_2}{\overset{}{CH}}-C_6H_4OMe-2 \right] \xrightarrow{H_2O}$$

$$\underset{\phi}{\overset{HOCH_2}{>}}=CHCH_2CH(OH)C_6H_4OMe-2 \xrightarrow[-H_2O]{H^\oplus} \underset{\phi}{\bigcirc}\overset{O}{\diagup}C_6H_4OMe-2$$

$$(XVIII) \qquad\qquad (XIX)$$

BETAINE-LIKE METHYLENE TRIPHENYLPHOSPHORANE

Chalcone undergoes Wittig reaction[65] to a limited extent (18%), thus

$$\phi_3P{=}CH{-}C_6H_4SO_2C_6H_5 + R_1COR_2 \longrightarrow$$

$$\begin{array}{c} R_1 \\ R_2 \end{array}{>}C{=}CH{-}C_6H_4{-}SO_2C_6H_5 + \phi_3PO$$

$$R_1 = \phi; \; R_2 = styryl$$

NITRONE

C,N-Diphenylnitrone and N-methyl-C-phenylnitrone are reported[66] to undergo 1,3-dipolar cycloaddition with p-substituted chalcone to yield the corresponding isoxazolidine ring system.

NITROGEN HETEROCYCLES

Chalcones have been reacted with several nitrogen heterocycles, for example, piperidine,[67] morpholine,[67] N-bromomorpholine,[68] piperizine,[69] indoles,[70,71] and azoles[72] to yield the corresponding Michael adducts. The following serve as examples[67]:

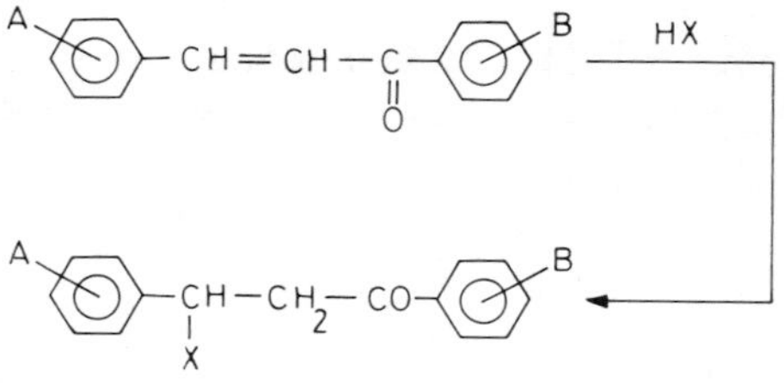

(A and B represent appropriate substituents and X stands for a pyridyl or morpholinyl residue.) The addition compounds are unstable and break up into its constituents in hot water.

A Michael adduct is formed by the interaction of chalcone with 3,5-dimethyl-4-nitroisoxazole. The adduct on reductive cyclization is reported to yield the corresponding azepine derivative.[73]

HYDRAZOIC ACID

The Schmidt reaction on highly hindered cis-chalcone is reported to yield quinoline derivatives.[74,75]

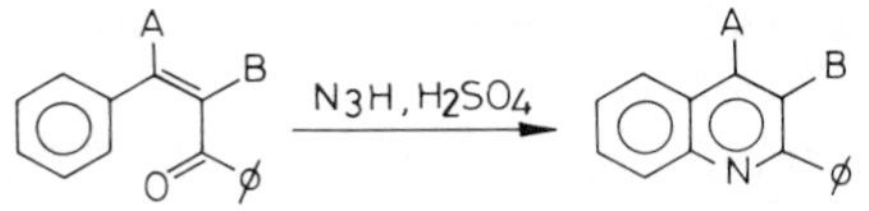

(XX) A = B = φ ;

(XXI) A = B = Br

The kinetics of this reaction has been studied.[76,77]

The reaction of hydrazoic acid with 2'-hydroxychalcone has been carried out.[78] The following heterocyclic compounds (**XXII–XXVI**) have been isolated[78]:

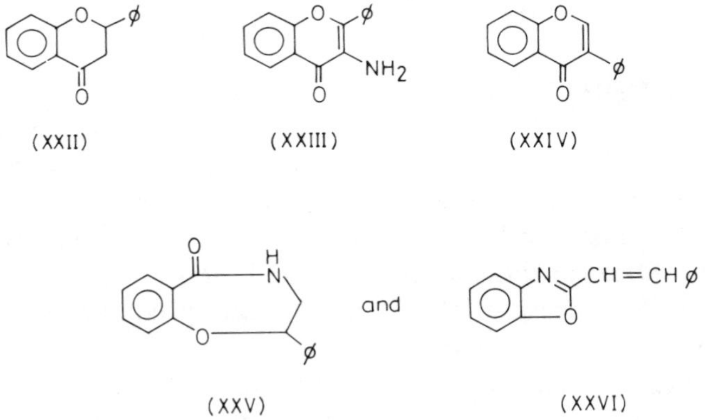

(XXII)   (XXIII)   (XXIV)

(XXV)   and   (XXVI)

HYDROXYLAMINE

Chalcone and substituted chalcones react with hydroxylamine hydrochloride to form the corresponding unsaturated ketoximes.[79-83] The ketoximes are reported to undergo a variety of reactions, cyclizations, catalytic hydrogenation, and Beckmann rearrangement.[79,80,90-93]

3,5-Diphenylisoxazoline is formed when chalcone is allowed to react with hydroxylamine in an alkaline medium.[79,80] It is believed that chalcone *syn*-oxime is formed in the reaction, which undergoes cyclization to the corresponding isoxazoline derivative.[81] Chalcones carrying substituents in the 4- and 4'-positions react, under acid conditions, with hydroxylamine hydrochloride to yield isoxazolines and chalcone *syn*-oximes. On the other hand, isoxazolines, along with other products, are formed under alkaline conditions.[84,85] Isoxazolines (**XXVII**) derived from naphthalene analogues of chalcone have been secured by their reaction, at a higher temperature, with hydroxylamine in presence of pyridine.[86]

In some cases the formation of isoxazole[82,88,89] and dihydroisoxazole[87] have been reported. The mechanism of isoxazole formation is reported to take place by 1,2-addition.[88]

HYDRAZINE

Hydrazones[94] or pyrazolines[95] are formed by the reaction of chalcone with hydrazine. 4,5-Hydro-1H-pyrazoles[87] are, however, produced from 2-hydroxychalcones in the above reaction.

PHENYLHYDRAZINE

Chalcones (**XXVIIIa**) react with phenylhydrazine[86,94,96,97] in acetic acid medium to yield the corresponding phenylhydrazones[96,98] (**XXIX**). The phenylhydrazones can be transformed into 1,3,5-triphenylpyrazolines (**XXX**) by refluxing these with acetic acid. Some chalcone phenylhydrazones (**XXVIIIb**) are labile and are readily converted to pyrazolines,[96] even at room temperature. If a mixture of chalcone and phenylhydrazine are reacted at higher temperature, the intermediate hydrazone undergoes cyclization to the corresponding pyrazoline.[86,97]

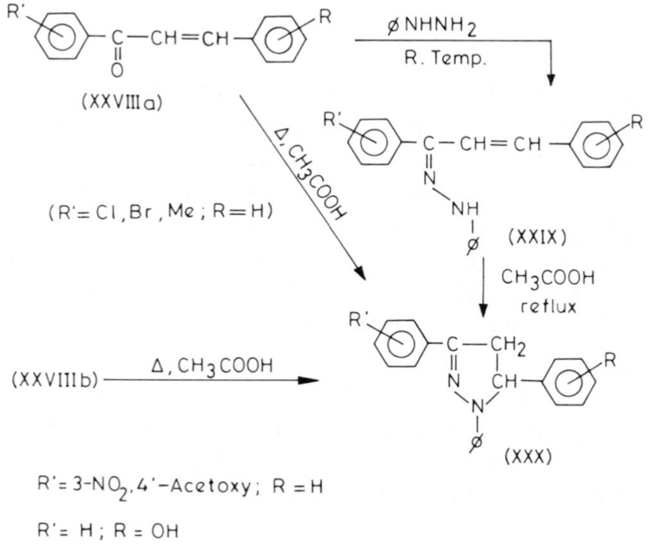

$R' = 3\text{-}NO_2, 4'\text{-}Acetoxy; \ R = H$

$R' = H; \ R = OH$

A qualitative test to distinguish phenylhydrazones from their isomeric pyrazolines has been reported.[96]

2,4-DINITROPHENYLHYDRAZINE (DNPH)[94,99–103]

The chief advantage of DNPH over phenylhydrazine as a reagent for characterization of chalcones is the ease of separation of the crystalline hydrazones with characteristic sharp melting points. The preparation of some substituted chalcone 2,4-dinitrophenylhydrazones are reported in the literature.[99–101]

The reactivity of carbonyl group in chalcones, that is, their formation of 2,4-dinitrophenylhydrazones, has been examined.[102,103] The close proximity of the hydroxyl group hinders the reactivity of the carbonyl group because of chelate formation. Substituents located in the *para*-position with respect to the carbonyl function alter, depending on their inductive effects,[102,103] the rate of 2,4-dinitrophenylhydrazone formation.

HYDRAZINE DERIVATIVES

Several other hydrazine derivatives,[104–107] *l*-menthydrazide,[104] 3,5-dinitro-4-tolylhydrazine,[105] phenylsulfonylhydrazide,[106] and tosylhydrazine,[107] have been used for the characterization of chalcones. One of these derivatives, tosylhydrazone, has an important synthetic application and is used as a starting material for the preparation of substituted cyclopropenes.[107]

DISUBSTITUTED PHOSPHINE OXIDE

Addition across the olefinic linkage occurs when chalcone reacts with disubstituted phosphine oxide,[108] under basic conditions, thus

$$(\phi CH_2)_2 P(O)H + \phi CH{=}CHCO\phi \xrightarrow{\text{EtOH–EtONa}}$$

$$(\phi CH_2)_2 P(O){-}\underset{\underset{\phi}{|}}{CH}{-}CH_2{-}CO\phi$$

$$(92\%)$$

N-BROMOSUCCINIMIDE (NBS)[109]

β-Phenylchalcone on treatment with *N*-bromosuccinimide gives the α-bromo derivative (**XXXI**). The reaction involves a radical mechanism and is initiated by a trace of Br_2 formed by the decomposition of *N*-bromosuccinimide.

$$\phi_2C=CH-\overset{\overset{\displaystyle O}{\|}}{C}-\phi \ + \ Br_2 \longrightarrow \phi_2C=\overset{\overset{\displaystyle Br}{|}}{C}-\overset{\overset{\displaystyle O}{\|}}{C}-\phi \ + \ HBr$$

$$(XXXI)$$

$$HBr \ + \ \begin{array}{c} CH_2-C \\ | \qquad \diagdown \\ CH_2-C \nearrow \end{array} N-Br \longrightarrow \begin{array}{c} CH_2-C \\ | \qquad \diagdown \\ CH_2-C \nearrow \end{array} NH \ + \ Br_2$$

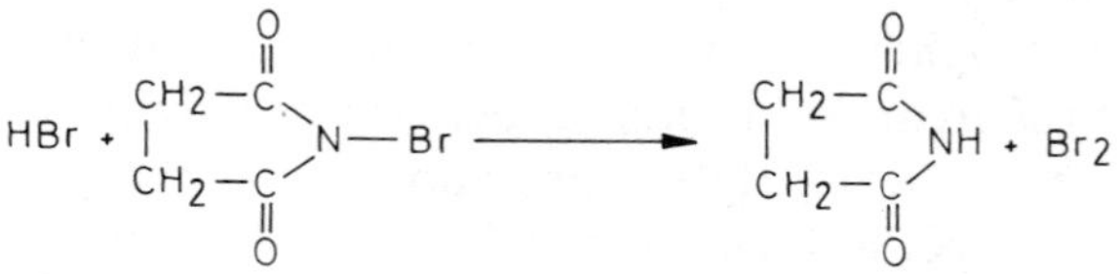

Bromoacetoxylation of several chalcones are reported and is carried out by the reaction of chalcone with NBS in acetic acid.[110]

$$\phi CH=CHCO\phi \ \xrightarrow{\ NBS \ + \ CH_3COOH \ }$$

$$\phi CH(OCOCH_3)CHBrCO\phi$$
$$+ \ \phi CHBrCH(OCOCH_3)CO\phi$$

Apparently α-bromo-β-acetoxy and β-bromo-α-acetoxy isomers are produced in the reaction, the relative proportion of which varies from chalcone to chalcone.

BROMINE AZIDE[111]

Bromine azide adds very slowly to chalcone in methylene chloride–nitromethane, in the following way:

$$\phi CH=CHCO\phi \ \xrightarrow{BrN_3} \ \phi CH(N_3)-CHBr-CO\phi$$

The reaction can, however, be made to go at a reasonable rate by incorporating an acid catalyst.

PYROLYSIS

Pyrolysis of chalcone at about 700° in the presence of aluminum bronze powder is reported to yield stilbene and 1,4-(*m*-biphenylyl)benzene.[112]

AUTOCONDENSATION

Chalcones are reported to undergo autocondensation in the presence of ethyl formate–perchloric acid to yield 2,4,6-triarylpyrylium salt in a poor yield (18–48%).[113]

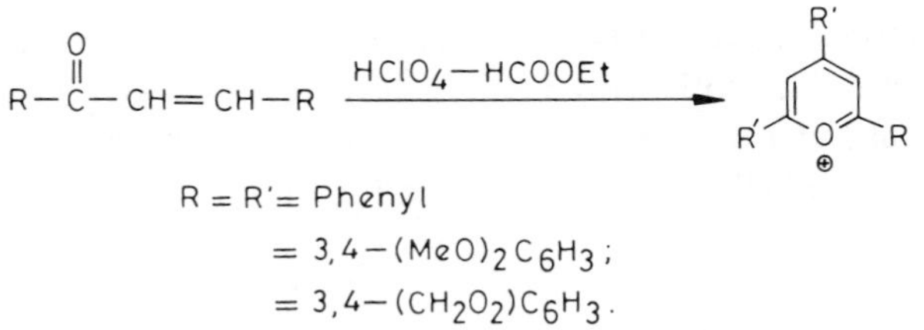

$$R-\overset{\overset{\displaystyle O}{\|}}{C}-CH=CH-R \xrightarrow{\text{HClO}_4-\text{HCOOEt}}$$

$$R = R' = \text{Phenyl}$$
$$= 3,4-(MeO)_2 C_6H_3 ;$$
$$= 3,4-(CH_2O_2)C_6H_3 .$$

COPOLYMERIZATION OF CHALCONES/CHALCONE ANALOGUES

Styrene

Substituted chalcone and styrene[114-116] form a copolymer (85% conversion[115]) involving the use of a free radical initiator. The product is described as hard, clear solid polymer, with a heat distortion point of 97° (cf. polystyrene, 78°). A resinous solid, however, is obtained by the ionic interpolymerization of chalcone and styrene with boron trifluoride[117] (at −80°) in methylene chloride.

Butadiene

1,3-Butadiene[115] monomer copolymerizes readily with α,β-unsaturated carbonyl compounds in the presence of benzoyl peroxide. The carbonyl compounds studied include, chalcone,[115] 2-chlorochalcone,[118] furfuralacetophenone,[118] furfural-*p*-chloroacetophenone,[118] and some

pyridine analogues[119] of chalcone, viz., 2-(3- and 4-) pyridalacetophen-
ones, 2-pyridal-4-chloroacetophenone, and 2-pyridal-2'-acetylaceto-
phenone. It may, however, be noted that 2-pyridal-4'-amino-
acetophenone[119] fails to form a copolymer with 1,3-butadiene.

Acrylonitrile and Isoprene[120]

Copolymers derived from acrylonitrile (and isoprene) and pyridine
analogues of chalcones are described in the literature.[120]

SYNTHESIS OF m-POLYPHENYLS[121]

Chalcone serves as a starting material for the synthesis of m-poly-
phenyl system. The following sequence of reaction steps are necessary
in bringing about the required transformation, taking the synthesis of
m-terphenyl as an illustrative example[121]:

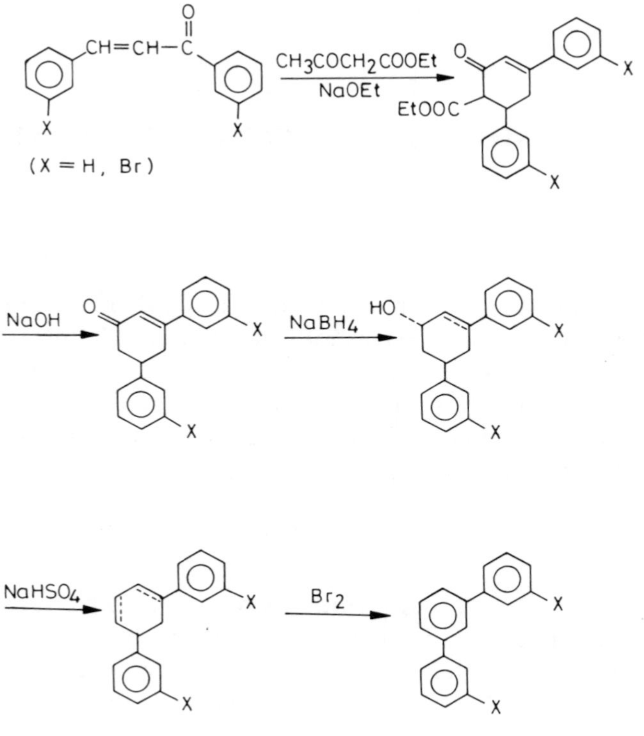

SYNTHESIS OF 1,3-DIPHENYLGLYCEROL[122]

Starting with chalcone and taking it through a series of chemical transformations the two isomers, *erythro–erythro* and *erythro–threo,* of 1,3-diphenylglycerol have been isolated[122] and identified.

SYNTHESIS OF ARYLBENZYLETHANOLAMINES

The title compounds have been obtained[123] by the following series of reactions. Thus

$$\phi COCH{=}CH\phi \quad \xrightarrow[\text{(H}_2)]{\text{Raney nickel}} \quad \phi COCH_2CH_2\phi \quad \xrightarrow{\text{HCl, MeONO}}$$

$$\phi COCH(NO)CH_2\phi \quad \xrightarrow[\text{ii)} \quad \text{HCl}]{\text{i) Raney nickel + NaOH}}$$

$$\phi CH(OH)(NH_2{\cdot}HCl)CH_2\phi \quad \xrightarrow{\text{NH}_3} \quad \phi CH(OH)NH_2CH_2\phi$$

REFERENCES

1 Shriner, R. L., and Kurosaws, T., *J. Am. Chem. Soc.*, **52**, 2538 (1930).

2 Carsky, P., Zuman, P., and Horak, V., *Coll. Szech. Chem. Commun.*, **30**, 4316 (1965).

3 Walker, E. A., and Young, J. R., *J. Chem. Soc.*, 2045 (1957).

4 Van Senden, K. G., and Koningsberger, C., *Tetrahedron*, **22**, 1301 (1966).

5 DeBlic, A., and Maroni, P., *Bull. Soc. Chim. France* (3–4), 512 (1975).

6 Wacek, A. V., and David, E., *Chem. Ber.*, **70B**, 190 (1937).

7 Pfeiffer, P., and Negreanu, P. A., *Chem. Ber.*, **50**, 1465 (1917).

8 Gillespie, R. J., and Leisten, J. A., *J. Chem. Soc.*, 1 (1954).

9 Allen, J. M., Johnston, K. M., Jones, J. F., and Shotter, R. G., *Tetrahedron*, **33**, 2083 (1977).

10 Hauser, C. R., and Tetenbaum, M. T., *J. Org. Chem.*, **23**, 1146 (1958).

11 Kratzl, K., and Daubner, H., *Chem. Ber.*, **77B**, 519 (1944).

12 Fromm, E., and Hubert, E., *Annalen*, **394**, 290.

13 Padwa, A., and Gruber, R., *Chem. Commun.*, (1), 5 (1969).

14 Rao, K. V., *J. Heterocycl. Chem.*, **12**, 725 (1975).

15 Bruce, D. B., Sorrie, A. J. S., and Thomson, R. H., *J. Chem. Soc.*, 2403 (1953).

16 Sammour, A., and Elkasaby, M., *U.A.R. J. Chem.*, **13**, 409 (1971); *Chem. Abstr.*, **177**, 101261f (1972).

17   Koelsch, C. F., *J. Org. Chem.*, **26**, 2590 (1961).

18   Davey, W., and Gwilt, J. R., *J. Chem. Soc.*, 1017 (1957).

19   Joshi, K. C., and Jauhar, A. K., *J. Indian Chem. Soc.*, **43**, 368 (1966).

20   Takatori, M., and Kanomate, K., *Bull. Chem. Soc. (Japan)*, **48**, 3411 (1975).

21   Narasimhachari, N., and Seshadri, T. R., *Proc. Indian Acad. Sci.*, **29A**, 265 (1949).

22   Dippy, J. F. J., and Palluel, A. L. L., *J. Chem. Soc.*, 1415 (1951).

23   Perrot, R., *Compt. Rend.*, **203**, 329 (1936).

24   Romano, C., *Ann. Chim. (Rome)*, **60**, 405 (1970); *Chem. Abstr.*, **73**, 87603c (1970).

25   Vorlander, D., and Eichwald, E., *Chem. Ber.*, **56B**, 1150 (1923).

26   Bloodworth, A. J., and Bruce, R. J., *J. Chem. Soc. C*, 1453 (1971).

27   Patnaik, A. K., Nayak, P. L., and Rout, M. K., *Indian J. Chem.*, **8**, 722 (1970).

28   Nesmeyanov, A. N., Shul'pin, G. B., Rybin, L. V., Gubenko, N. T., Rybinskaya, M. I., and Petrovsky, P. V. and Robas, V. I., *Zh. Obshch. Khim.*, **44**, 2032 (1974).

29   Nesmeyanov, A. N., Rybin, L. V., Gubenko, N. T., Petrovsky, P. V., and Rybinskaya, M. I., *Zh. Obshch. Khim.*, **42**, 2473 (1972).

30   Cabak, J., and Veres, K., *Radiochem. Radioanal. Letters* (12), 6, 331 (1972).

31   Kursanov, D. N., and Parnes, Z. N., *Dok. Akad. Nauk SSSR*, **91**, 1125 (1953); *Chem. Abstr.*, **48**, 10549c (1954).

32   Zinn, M. F., Harris, T. M., Hill, D. G., and Hauser, C. R., *J. Am. Chem. Soc.*, **85**, 71 (1963).

33   Yamamura, K., *J. Org. Chem.*, **43**, 724 (1978).

34   Worral, D. E., and Bradway, C. J., *J. Am. Chem. Soc.*, **58**, 1607 (1936).

35   Fishman, N., and Zuffanti, S., *J. Am. Chem. Soc.*, **73**, 4466 (1951).

36   Davey, W., and Tivey, D. J., *J. Chem. Soc.*, 2276 (1958).

37   Colonna, S., Hiemstra, H., and Wynberg, H., *Chem. Commun.* (6), 238 (1978).

38   Profft, E., and Wolf, E., *J. Prakt. Chem.*, **19**, 192 (1963).

39   Seter, J., *Israel J. Chem.*, **4**, 1 (1966); *Chem. Abstr.*, **66**, 2294r (1967).

40   Solomonovici, A., *Israel J. Technol.*, **7**, 441 (1969); *Chem. Abstr.*, **72**, 89961w (1970).

41   Smith, L. I., and Pings, W. B., *J. Org. Chem.*, **2**, 23 (1937).

42   Ghate, C. G., Kaushal, K., and Deshpande, S. S., *J. Indian Chem. Soc.*, **27**, 633 (1950).

43   Johnson, W. S., and Birkeland, S. P., *Tetrahedron Letters* (5), 1 (1960).

44   Myers, P. L., and Lewis, J. W., *J. Heterocycl. Chem.*, **10**, 165 (1973).

45   Rumyantseva, K. S., Rumyantsev, N. P., Chekmaeva, G. E., and Kirlyanova, R. P., *Uch. Zap. Mord. Univ.*, No. 81, 75 (1971); *Chem. Abstr.*, **78**, 57948e (1973).

46   Sasson, Y., Cohen, M., and Blume, J., *Synthesis*, **6**, 359 (1973).

47 Dedieu, M., and Pascal, Y. L., *C. R. Acad. Sci., Ser. C,* **278,** 1425 (1974); *Chem. Abstr.,* **82,** 42653p (1975).

48 Gilman, H., Fullhart, L., and Cason, L. F., *J. Org. Chem.,* **21,** 826 (1956).

49 Davey, W. A., and Gwilt, J. R., *J. Chem. Soc.,* 1015 (1957).

50 Omote, Y., Komatsu, T., Kobayashi, R., and Sugiyama, N., *Nippon Kagaku Kaishi* (4), 780 (1972); *Chem. Abstr.,* **77,** 33637f (1972).

51 Hankovszky, O. H., Hideg, K., and Llyod, D., *J. Chem. Soc. Perkin. Trans.,* **1,** 1619 (1974).

52 Gilman, H., and Cason, L. F., *J. Am. Chem. Soc.,* **73,** 1074 (1951).

53 Ferreira, D., and Roux, D. G., *J. Chem. Soc. Perkin. Trans. I,* 134 (1977).

54 Lévai, A., and Bognár, R., *Acta Chim.,* **92,** 415 (1977).

55 Staudinger, H., and Karlsruhe, R. E., *Annalen,* **401,** 263.

56 McElvain, S. M., and Cohen, H., *J. Am. Chem. Soc.,* **64,** 260 (1942).

57 Saigo, K., Osaki, M., and Mukaiyama, T., *Chem. Letters* (2), 163 (1976).

58 Barker, S. A., and Riley, T., *J. Chem. Soc. Perkin Trans. I* (6), 809 (1972).

59 Connor, R., Flemming, C. L., Jr., and Clayton, T., *J. Am. Chem. Soc.,* **58,** 1386 (1936).

60 Ukai, S., Hirose, K., and Kayano, M., *Yakugaku Zasshi,* **95,** 299 (1975); *Chem. Abstr.,* **83,** 43125a (1975).

61 Gautier, J. A., Miocque, M., Moskowitz, H., and Blac-Guenee, J., French Patent 2,031,714 (1970); *Chem. Abstr.,* **75,** 48691b (1971).

62 Corey, E. J., and Chaykovsky, M., *J. Am. Chem. Soc.,* **84,** 3782 (1962).

63 Claude, A., *C. R. Acad. Sci., Paris, Ser. C,* **264,** 1128 (1967).

64 Donnelly, J. A., Bennett, P., O'Brien, S., and O'Grady, J., *Chem. Ind. (London),* 500 (1972).

65 Siemiatycki, M. S., Carretto, J., and Malbee, F., *Bull. Soc. Chim. France,* 125 (1962).

66 Iwakura, Y., Uno, K., Kihara, Y., Setsu, M., and Yamamoto, S., *Nippon Kagaku Kaishi* (8), 1448 (1972); *Chem. Abstr.,* **77,** 139864t (1972).

67 Kozlov, N. S., and Kozlov, G. N., *Zh. Obshch. Khim.,* **33,** 2184 (1963).

68 Southwick, P. L., and Walsh, W. L., *J. Am. Chem. Soc.,* **77,** 405 (1955).

69 Stewart, V. E., and Pollard, C. B., *J. Am. Chem. Soc.,* **58,** 1980 (1936).

70 Merchant, J. R., Salagar, S. S., and Chhatriwala, K. M., *Curr. Sci. (India),* **35,** 123 (1966).

71 Merchant, J. R., Chhatriwala, K. M., Salagar, S. S., and Patell, J. R., *J. Indian Chem. Soc.,* **48,** 613 (1971).

72 Moffat, J., *J. Am. Chem. Soc.,* **77,** 2572 (1955).

73 Rao, C. J., and Murthy, A. K., *Indian J. Chem.,* **16B,** 636 (1978).

74 Pratt, R. E., *Diss. Abstr.,* **27B,** 2656 (1967).

75 Pratt, R. E., Welstead, W. J., Jr., and Lutz, R. E., *J. Heterocycl. Chem.,* **7,** 1051 (1970).

76   Mirek, J., *Bull. Acad. Polon. Sci., Ser. Sci. Chim.*, **10**, 421 (1962); *Chem. Abstr.*, **59**, 6226g (1963).

77   Mirek, J., *Zesz. Nauk. Uniw. Jagiellon., Pr. Chem.*, No. 10, 61 (1965); *Chem. Abstr.*, **66**, 37125h (1967).

78   Misiti, D., *Ann. 1st Super. Sanita*, **9** Part 2–3, 174 (1973); *Chem. Abstr.*, **81**, 63599v (1974).

79   Auwers, K. V., and Seyfried, M., *Annalen*, **484**, 178 (1930).

80   Auwers, K. V., and Muller, H., *J. Prakt. Chem.*, **137**, 57 (1933).

81   Meisenheimer, J., and Campbell, N., *Annalen*, **539**, 93 (1939).

82   Samula, K., *Roczniki Chem.*, **45**, 2063 (1971); *Chem. Abstr.*, **76**, 153509s (1972).

83   Henrich, F., *Annalen*, **351**, 172.

84   Unterhalt, B., *Pharm. Zentralhalle*, **107**, 356 (1968); *Chem. Abstr.*, **69**, 99459r (1968).

85   Uterhalt, B., *Pharm. Zentralhalle*, **107**, 356 (1968); *Chem. Abstr.*, **70**, 57351q (1969).

86   Sammour, A. E. A., *Tetrahedron*, **20**, 1067 (1964).

87   Jurd, L., *Tetrahedron*, **31**, 2884 (1975).

88   Blecker, H. H., *Diss. Abstr.*, **15**, 2407 (1955).

89   Shenoi, B. B., Shah, R. C., and Wheeler, T. S., *J. Chem. Soc.*, 247 (1940).

90   Merz, K. W., *Chem. Ber.*, **63B**, 2951 (1930).

91   Dornow, A., and Frese, A., *Arch. Pharm.*, **285**, 463 (1952); *Chem. Abstr.*, **48**, 11374a (1954).

92   Ruchkin, V. E., Shvetsova-Shilovskaya, K. D., and Mel'nikov, N. N., *Probl. Poluch. Poluprod. Prom. Org. Sin., Akad Nauk SSSR, Otd. Obshch. Tekh. Khim.*, 66 (1967); *Chem. Abstr.*, **68**, 95417g (1968).

93   Lenka, S., Nayak, P. L., and Rout, M. K., *J. Inst. Chem. (India)*, **42**, 162 (1970).

94   Kallay, F., and Janzso, G., *Kem. Kozl.*, **42**, 213 (1974); *Chem. Abstr.*, **83**, 9701a (1975).

95   Sammour, A., Selim, M. I. B., and Sayed, G. H., *U.A.R. J. Chem.*, **14**, 235 (1971); *Chem. Abstr.*, **77**, 139878a (1972).

96   Raiford, L. C., and Peterson, W. J., *J. Org. Chem.*, **1**, 544 (1937).

97   Hanson, G. A., *Bull. Soc. Chim. Belges*, **67**, 707 (1958); *Chem. Abstr.*, **53**, 18007a (1959).

98   Raiford, L. C., and Davis, H. L., *J. Am. Chem. Soc.*, **50**, 156 (1927).

99   Dhar, D. N., *J. Indian Chem. Soc.*, **37**, 320 (1960).

100   Allen, C. F. H., *J. Am. Chem. Soc.*, **52**, 2955 (1930).

101   Langlais, M., Buzas, A., Soussan, G., and Freon, P., *Compt. Rend.*, **261**, 2920 (1965).

102   David, E. R., and Bognár, R., *Acta Chim. Acad. Sci. Hung.*, **84**, 335 (1975); *Chem. Abstr.*, **82**, 169686g (1975).

103   David, E. R., Szabo, G. B., Korponai, K., and Bognár, R., *Magy. Kem. Foly.*, **81**, 328 (1975); *Chem. Abstr.*, **83**, 177721a (1975).

104 Woodward, R. B., Kohman, T. P., and Harris, G. C., *J. Am. Chem. Soc.*, **63**, 120 (1941).

105 Sah, P. P. T., *J. Chinese Chem. Soc.*, **14**, 45 (1946); *Chem. Abstr.*, **43**, 6972e (1949).

106 Dornow, A., and Bartsch, W., *Annalen*, **602**, 23 (1957).

107 Duerr, H., *Angew. Chem. Inter. Ed.*, **6**, 1804 (1967).

108 Miller, R. C., Bradley, J. S., and Hamilton, L. A., *J. Am. Chem. Soc.*, **78**, 5299 (1956).

109 Southwick, P. L., Pursglove, L. A., and Numerof, P., *J. Am. Chem. Soc.*, **72**, 1600 (1950).

110 Iovchev, A., Spasov, S. L., Stefanovskii, Y. N., Stoilov, L., and Gocheva, V., *Monatsh. Chem.*, **100**, 51 (1969).

111 Hassner, A., and Boerwinkle, F., *J. Am. Chem. Soc.*, **90**, 216 (1968).

112 Dhar, D. N., and Munjal, R. C., *Indian J. Chem.*, **12**, 1014 (1974).

113 Dorofeenko, G. N., Olekhnovich, E. P., and Laukhina, L. I., *Zh. Organ. Khim.*, **7**, 1296 (1971); *Chem. Abstr.*, **75**, 98392a (1974).

114 Chapin, E. C., U.S. Patent 2,496,697 (1950); *Chem. Abstr.*, **44**, 4723c (1950).

115 Marvel, C. S., McCorkle, J. E., Fukuto, T. R., and Wright, J. C., *J. Polymer. Sci.*, **6**, 776 (1951).

116 Marvel, C. S., and Wright, J. C., *J. Polymer. Sci.*, **8**, 495 (1952).

117 Coleman, L. E., Jr., and Jones, F. B., *J. Polymer. Sci.*, **31**, 528 (1958).

118 Marvel, C. S., Peterson, W. R., Inskip, H. K., McCorkle, J. E., Taft, W. K., and Labbe, B. G., *Ind. Eng. Chem.*, **45**, 1532 (1958).

119 Marvel, C. S., Coleman, L. E., Jr., Scott, G. P., Taft, W. K., and Labbe, B. G., *Ind. Eng. Chem.*, **48**, 214 (1956).

120 Marvel, C. S., Coleman, L. E., Jr., and Scott, G. P., *J. Org. Chem.*, **20**, 1785 (1955).

121 Sunshine, N. B., and Woods, G. F., *J. Org. Chem.*, **28**, 2517 (1973).

122 Sasaki, S., *Nippon Kagaku Zasshi*, **80**, 531 (1959); *Chem. Abstr.*, **55**, 4422e (1961).

123 Bar, D., and Debruyre, Mme. E., *Ann. Pharm. Franc.*, **16**, 235 (1958).

Chapter Sixteen

Photochemistry of Chalcones and Their Derivatives

PHOTOCHEMISTRY OF CHALCONES

Photoisomerization of *trans*-chalcone,[1] *trans*-2-hydroxychalcone[2] and other substituted chalcones,[3,4] and heterocyclic analogues of chalcones[5] into their corresponding *cis* isomers has been described in the literature. *cis*-2-Methoxychalcone is convertible to *cis*-2-hydroxychalcone (photochemical demethylation)[2] by prolonged irradiation to sunlight. Some chalcones, on the other hand, have a great tendency to undergo resinification[6] when irradiated either in the solid state or solution, viz., 4,4'-dimethyl- (and 4,4'-dimethoxy-) chalcones. The solid-state photochemistry of some 2'-nitrochalcones has been studied.[7] Various parameters seem to govern the specific pathway followed by the photochemical reaction, for example, molecular conformation and its retention and the molecular packings.

PHOTOCHEMICAL DIMERIZATION

Photochemical dimerization is reported to occur in the case of several chalcones: chalcone,[6,8] 4'-methylchalcone,[6] 4-methoxychalcone,[10] and the thiophene analogue[11] of chalcone. Dimerization, however, does not take place with chalcone or 4-methoxychalcone if these are irradiated in presence of uranyl chloride.[12] The cyclobutane type of structure[6,8] (see below) has been assigned to chalcone dimers. These results have been arrived at on the basis of data obtained by physical[8,9] and chemical methods.[6,10]

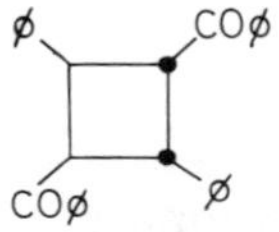

(m.p. 226°) 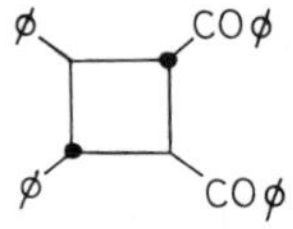   (m.p. 126°)

Photoinduced dimerization of 4-methoxychalcone is reported[13] to take place with the aid of 9,10-dihydroanthracene. Two types of dimerides,[13] head–tail and head–head orientations, are produced when the photoreaction is carried out in ethanol and acetonitrile, respectively. Chalcone dimers of the following structure have been obtained[14] in high yield by the photolysis of a solution of chalcone under appropriate conditions[15]:

$$X—C_6H_4—CH_2—CH—CO—\phi$$
$$|$$
$$X—C_6H_4—CH_2—CH—CO—\phi$$
$$X = H, \text{4-OMe, or 4-Cl}$$

Photolysis of 2-hydroxychalcone in ethanol is reported to yield 2-ethoxy-flav-3-ene (96%) and a very small amount (1%) of flavone, according to the reaction[16] :

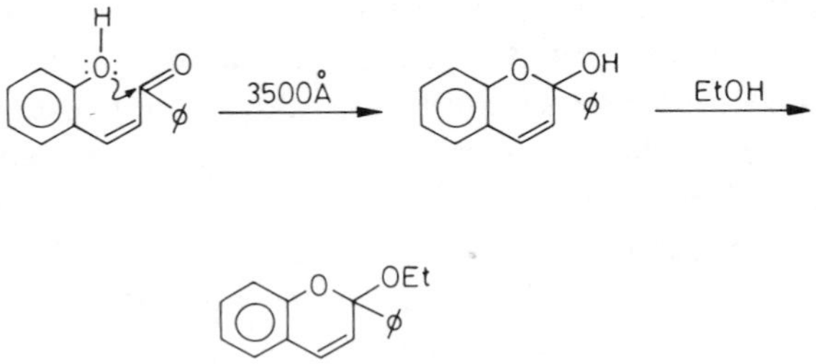

Methylene blue-sensitized photochemical oxidation by visible light has been reported[17] in the case of 2'-hydroxy-4',6',3,4-tetramethoxy-chalcone leading to the formation of 5,7,3',4'-tetramethoxyflavanonol. The 3-hydroxylic function in the flavanonol is assumed to arise from the hydroxylic radical generated by the photolysis of aqueous methanol, used as the solvent.

PHOTOCHEMISTRY OF CHALCONE DERIVATIVES

Flavanone

UV irradiation of flavanone is reported[18] to yield three products, 2'-hydroxychalcone (20%), 4-phenyl-dihydrocoumarin (13%), and salicylic acid (4%). The mechanism of the reaction is as follows[18]:

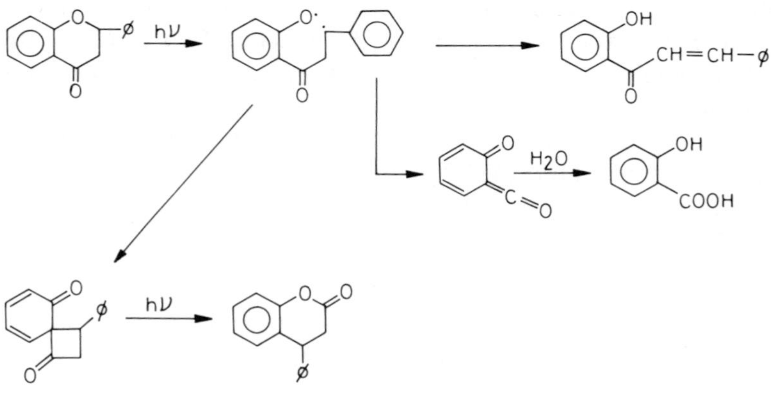

Chalcone Epoxides

Chalcone epoxides are reported to undergo photooxidative cleavage, yielding a mixture of acid and aldehyde.[19] Thus, for example, 2',3,4,4'-tetramethoxychalcone epoxide[19] yields veratraldehyde and 2,4-dimethoxybenzoic acid under these conditions. According to another report *trans*-chalcone epoxide and *trans*-2-methoxychalcone epoxide on irradiation yield dibenzoylmethanes, *trans–cis* isomerization, and fragmentation via acylcarbene intermediates.[20] Evidence has been put forward about the intermediacy of dioxoles when substituted chalcone epoxides are photolyzed.[21] The dioxoles are produced by the cleavage of the C_α—C_β oxirane bond, leading to the formation of carbonyl ylides, which subsequently rearrange to dioxoles.

Chalcone Semicarbazones

Some of the chalcone semicarbazones are reported to exhibit noticeable photochemical isomerism.[22] Examples are *p*-methylchalcone and *p*-methoxy-*p'*-methylchalcone.[22]

REFERENCES

1 Tryon, M., and Wall, L. A., *J. Phys. Chem.*, **62**, 697 (1958).

2 Jurd, L., *Tetrahedron*, **25**, 2367 (1969).

3 Ferreira, D., and Roux, D. G., *J. Chem. Soc. Perkin Trans. I*, No. 2, 134 (1977).

4 Zahir, S. A., *J. Appl. Polymer Sci.*, **23**, 1355 (1979).

5 Reinhardt, M., Mitina, V. G., Zadorozhnyi, B. A., and Lavrushin, V. F., *Zh. Obshch. Khim.*, **48**, 2759 (1978).

6 Stobbe, H., and Bremer, K. *J. Prakt. Chem.*, **123**, 1 (1929).

7 Jungk, A. E., Luwisch, M., Princhar, S., and Schmidt, G. M. J., *Isr. J. Chem.*, **16**, 308 (1977).

8 Montaudo, G., and Caccamese, S., *J. Org. Chem.*, **38**, 710 (1973).

9 Caccamese, S., McMillan, J. A., and Montaudo, G., *J. Org. Chem.*, **43**, 2703 (1978).

10 Stobbe, H., and Hensel, A., *Chem. Ber.*, **59B**, 2254 (1926).

11 Wynberg, H., Groen, M. B., and Kellogg, R. M., *J. Org. Chem.*, **35**, 2828 (1970).

12 Praetorius, P., and Korn, F., *Chem. Ber.*, **43**, 2744 (1910).

13 Burton, G. W., *Diss. Abstr.*, **25**, 4400 (1965).

14 Sugden, J. K., *Synthetic Commun.*, **6**, 93 (1976).

15 Evans, P. G. E., Sugden, J. K., and Abbe, N. J. V., *Pharm. Acta Helv.*, **50**, 94 (1975).

16 Dewar, D., and Sutherland, R. G., *J. Chem. Soc., D*, 272 (1970).

17 Chawla, H. M., Chibber, S. S., and Sharma, A., *Tetrahedron Letters*, 2713 (1978).

18 Mack, P. O. L., and Pinhey, J. T., *Chem. Commun.* (8), 451 (1972).

19 Parthasarthy, M. R., and Sharma, D. K., *Indian J. Chem.*, **12**, 1009 (1974).

20 Dewar, D. J., and Sutherland, R. G., *J. Chem. Soc., Perkin Trans.*, **2**, 1522 (1977).

21 Lee, G. A., *J. Org. Chem.*, **43**, 4256 (1978).

22 Heilbron, I. M., and Wilson, F. J., *J. Chem. Soc.*, **103**, 1504 (1913).

Chapter Seventeen

Polarographic Studies of Chalcones, Chalcone Analogues, and Their Derivatives

POLAROGRAPHIC REDUCTION OF CHALCONE

Chalcone undergoes reduction on a dropping mercury electrode yielding a variety of products, depending on the conditions of the experiment. The present picture about the mechanism involved in the formation of these products from chalcone, for example, has emerged as a result of the work of several investigations[1-14,17-19] and is summarized below:

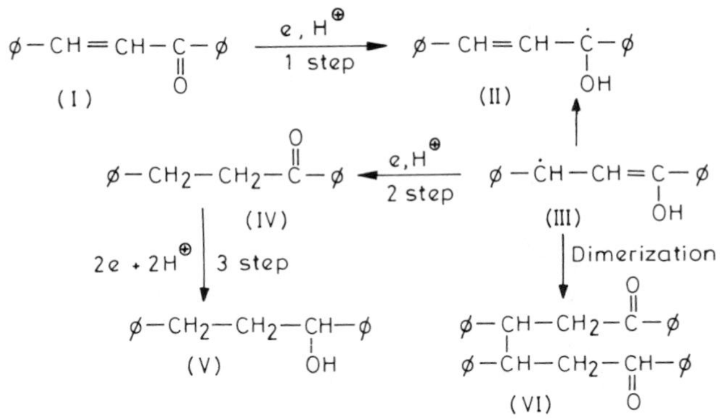

Mechanism

In the acid medium chalcone (**I**) undergoes electrode reduction to yield a saturated ketone[8] (**IV**). In the neutral or alkaline media, however, the reduction proceeds past the saturated ketone stage to yield the corresponding alcohol[8] (**V**). In the reaction (**I–IV**) the reduction of the ethylenic bond is a secondary process connected with a rearrangement of the free radical formed after the addition of one electron and one proton to the carbonyl group (**II–III**). The formation of other products has been interpreted to arise either by dimerization[6] (**VI**) or disproportionation of the radicals generated in the process.

POLAROGRAPHIC REDUCTION OF CHALCONE IN DIMETHYLFORMAMIDE (DMF)

Polarographic reduction of chalcone has been reported in anhydrous dimethylformamide in the presence and absence of carbon dioxide.[9] In the former case two acids, α-phenyl-β-benzoylpropionic acid, **XII** and

XIV, are formed. The mechanism of formation of **XII** can be rationalized as follows:[9]

$$\phi-CH{=}CHCO\,\phi \; + \; e \longrightarrow \phi\,\overset{\cdot}{C}H-CH{=}\underset{\underset{O^{\ominus}}{|}}{C}-\phi \longleftrightarrow$$

$$\text{(VII)}$$

$$\underset{\text{(VIII)}}{\phi-\overset{\cdot}{C}H-\underset{\ominus}{C}H-CO\,\phi} \xrightarrow{CO_2} \underset{\text{(IX)}}{\phi-\overset{\cdot}{C}H-\underset{\underset{COO^{\ominus}}{|}}{C}H-CO\,\phi} \xrightarrow{\;e\;}$$

$$\underset{\text{(X)}}{\phi-\underset{\ominus}{C}H-\underset{\underset{COO^{\ominus}}{|}}{C}H-CO\,\phi} \xrightarrow{CO_2} \underset{\text{(XI)}}{\phi-\underset{\underset{COO^{\ominus}}{|}}{C}H-\overset{\overset{COO^{\ominus}}{|}}{C}H-CO\,\phi}$$

$$\xrightarrow{H^{\oplus}} \underset{\text{(XII)}}{\phi-\underset{\underset{COOH}{|}}{C}H-CH_2-CO\,\phi}$$

Mechanism

Alternatively, the anion free radical **IX** may add to the double bond of another chalcone molecule and then be reduced further to yield finally the dimeric monocarboxylic acid **XVI**:

$$IX \; + \; \phi CH{=}CHCO\,\phi \longrightarrow \underset{\text{(XIII)}}{\overset{\displaystyle \phi-CH-\overset{\overset{COO^{\ominus}}{|}}{C}H-CO\,\phi}{\phi-\overset{\cdot}{C}H-CH-CO\,\phi}}$$

$$\xrightarrow{\;e\;} \underset{\text{(XIV)}}{\overset{\displaystyle \phi-CH-\overset{\overset{COO^{\ominus}}{|}}{C}H-CO\,\phi}{\phi-\underset{\ominus}{C}H-CH-CO\,\phi}} \xrightarrow{\text{Cyclisation}} \underset{\text{(XV)}}{\overset{\displaystyle \phi-CH-\overset{\overset{COO^{\ominus}}{|}}{C}H{\diagdown}_{C}{\diagup}^{O^{\ominus}}}{\phi-\underset{\underset{CO\,\phi}{|}}{C}H-CH{\diagup}^{\phi}}}$$

$$\xrightarrow{H^{\oplus}} \underset{\text{(XVI)}}{\phi{-}\overset{\overset{COOH}{}}{\underset{\phi}{|}}\overset{OH}{\underset{\phi}{}}CO\,\phi}$$

In the absence of carbon dioxide, the chalcone dianion furnishes the polymer[9] **XVII**. The polymer must have been produced by a series of Michael-type additions of the dianion intermediate to chalcone:

$$\phi-\overset{\ominus}{\underset{\ominus}{C}}H-CH=\overset{\overset{O}{\overset{|}{\overset{\ominus}{|}}}}{C}-\phi \;+\; \phi-CH=CH-CO\,\phi \longrightarrow$$

$$\begin{array}{c}\phi-CH-CH=\overset{\overset{O\;\ominus}{|}}{C}-\phi\\ |\\ \phi-CH-CH=\underset{\underset{O\;\ominus}{|}}{C}-\phi\end{array} \longleftrightarrow \begin{array}{c}\phi-CH-CH=\overset{\overset{O\;\ominus}{|}}{C}-\phi\\ |\\ \phi-\overset{}{C}H-CH-CO-\phi\\ \underset{\ominus}{}\end{array}$$

$$\xrightarrow[\phi\,CH=CHCO\,\phi]{} \begin{array}{c}\phi-CH-CH=\overset{\overset{O\;\ominus}{|}}{C}-\phi\\ \diagdown\\ \phi-CH-CH-CO-\phi \longrightarrow (\phi-\overset{|}{C}H-\overset{|}{C}H-CO\,\phi)_n\\ |\qquad\qquad\qquad\qquad (XVII)\\ \phi-\underset{\ominus}{C}H-CH-CO-\phi\end{array}$$

The formation of the unsaturated trimer **XVIII** in the polarographic reduction of chalcone (in the absence of carbon dioxide) has been assumed to involve the following intermediates[9]:

$$\begin{array}{c}\phi-CH-CH=\overset{\overset{O\;\ominus}{|}}{C}-\phi\\ |\\ \phi-CH-\underset{\cdot}{C}H-CO-\phi\end{array} \;+\; \phi\,CH=CH-CO\,\phi \longrightarrow$$

$$\begin{array}{c}\phi-CH-CH=\overset{\overset{O\;\ominus}{|}}{C}-\phi\\ |\\ \phi-\overset{\cdot}{C}H-CH-CO-\phi\\ |\\ \phi-\underset{\cdot}{C}H-\overset{}{C}H-CO-\phi\end{array} \xrightarrow{\text{Disproportionates}} \begin{array}{c}\phi-CHCH_2CO-\phi\\ |\\ \phi-CH-CHCO-\phi\,+\\ |\\ \phi-HC=\overset{}{C}-CO-\phi\\ (XVIII)\end{array}$$

$$\begin{array}{c}\phi-CH-CH_2CO-\phi\\ |\\ \phi-CH-CH-CO-\phi\\ |\\ \phi-\underset{2}{C}H-CH-CO-\phi\end{array}$$
$$\text{(Not isolated)}$$

SUBSTITUENT EFFECTS ON POLAROGRAPHIC REDUCTION OF CHALCONES

The ease of polarographic reduction is dependent on the nature of the substituents.[4,15] Thus chalcones containing a hydroxyl[4] or methoxy[4,15] function (in conjugation with the carbonyl group[4]) are less prone to reduction, as compared to chalcone bearing an acetoxy function. This effect is less pronounced in an acid medium than in a neutral medium.[16]

POLAROGRAPHIC REDUCTION OF CHALCONE ANALOGUES

The polarographic reduction of ferrocenyl[7,20] and heterocyclic analogues of chalcone (viz., pyrrolyl, furanyl, thienyl, and pyridyl) is reported in the literature. The polarographic reduction of furan chalcone is reported to take place via radical anion formation.[23,24] The behavior of pyridyl[21,22] analogues of chalcone in polarographic reduction is similar to that of chalcone.[25] In an alkaline medium other reactions[26] manifest themselves, for example, the hydration of the double bond, followed by cyclization to chromanones (pH $\sim$ 12.5), which are in turn polarographically reduced further.

The polarographic reduction behavior of chalcone has been studied and compared with its acetylenic analogue, 1,3-diphenyl-1-propyn-3-one.[27]

POLAROGRAPHIC ANALYSIS OF 2'-HYDROXYCHALCONE–FLAVANONE MIXTURE

In a weakly acidic or neutral medium chalcone shows two polarographic reduction potentials,[3] while flavanone has only one. Based on this property it is possible to determine 2'-hydroxychalcone in the presence of its isomer, flavanone.[3] The polarographic method could also be used to monitor the rate of formation of chalcone from its components.[3]

POLAROGRAPHIC CONVERSION OF 2'-HYDROXYCHALCONE TO CHROMANONE

A polarographic study of the transformation of *ortho*-hydroxychalcone to chromanones has been reported.[28,29] The percentage of hydroxychalcone at equilibrium is independent[29] of pH (range 5–9); however, with pH > 9, the equilibrium is displaced in favor of chalcone, attended with immediate decomposition.

POLAROGRAPHIC CONVERSION OF CHALCONE TO CHALCONE HYDRAZONE

A polarographic study has been reported of the reaction of chalcone with phenylhydrazine.[30] The conclusion[31] is that the first step is the reaction on mercury cathode of the carbonyl group of the chalcone.

POLAROGRAPHIC BEHAVIOR OF PYRIDINE ANALOGUE OF CHALCONE OXIME

The pyridyl analogue of chalcone oxime,[32] $R_1C(NOH)$—CH=CH—R_2 ($R_1 = R_2 = $ 3-pyridyl) is reported to undergo protonation and polarographic reduction at the oximo group and then at the olefinic linkage.

REFERENCES

1 Pasternak, R., and Halban, H. V., *Helv. Chim Acta*, **29**, 190 (1946).

2 Pasternak, R., *Helv. Chim. Acta*, **31**, 753 (1948).

3 Schraufstätter, E., *Experientia*, **4**, 192 (1948).

4 Geissman, T. A., and Friess, S. L., *J. Am. Chem. Soc.*, **71**, 3893 (1949).

5 Korshunov, I. A., and Vodzinskiĭ, Y. V., *Zh. Fiz. Khim.*, **27**, 1152 (1953); *Chem. Abstr.*, **48**, 5674c (1954).

6 Vodzinskiĭ, Y. V., and Korshunov, I. A., *Uchenye Zapiski Gor'kovsk, Gos. Univ. im N. I. Lobachevskogo, Ser. Khim.*, **32**, 25 (1958); *Chem. Abstr.*, **54**, 17113h (1960).

7 Tirouflet, J., Laviron, E., Metzger, J., and Boichard, J., *Coll. Czech. Chem. Commun.*, **25**, 3277 (1960).

8 Lavrushin, V. F., Bezuglyi, V. D., and Belous, G. G., *Zh. Obshch. Khim.*, **33**, 1711 (1963).

9   Wawzonek, S., and Gundersen, A., *J. Electrochem. Soc.*, **111**, 324 (1964).

10 Ryvolova-Kejharova, A., and Zuman, P., *J. Electroanal. Chem. Interfacial Electrochem.*, **21**, 197 (1969).

11 Ash, M. L., O'Brien, F. L., and Boykin, D. W., Jr., *J. Org. Chem.*, **37**, 106 (1972).

12 Seguin, J. P., Doucet, J. P., and Uzan, R., *C. Acad. Sci., Ser. C.*, **278**, 129 (1974); *Chem. Abstr.*, **80**, 95024z (1974).

13 Taran, L. A., Berezina, S. I., Smolentseva, L. G., and Kitaev, Y. P., *Izv. Akad. Nauk SSSR, Ser. Khim.* (11), 2611 (1974); *Chem. Abstr.*, **82**, 78456r (1975).

14 Katiyar, S. S., Lalithambika, M., and Dhar, D. N., J. Electroanal. *Chem. Interfacial Electrochem.*, **53**, 449 (1974).

15 Lavrushin, V. F., Bezuglyi, V. D., and Belous, G. G., *Zh. Obshch. Khim.*, **34**, 13 (1964).

16 Bezuglyi, V. D., Lavrushin, V. F., and Belous, G. G., *Zh. Ogshch. Khim.*, **35**, 606 (1965).

17 Yanovskaya, L. A., Umirzakov, B., and Kucherov, V. F., *Izv. Akad. Nauk SSSR, Ser. Khim.* (4), 823 (1972); *Chem. Abstr.*, **77**, 69365e (1972).

18 Bala, S., and Bannerjee, N. R., *Indian J. Chem.*, **15A**, 777 (1977).

19 Butkiewicz, K., *J. Electroanal. Chem. Interfacial Electrochem.*, **89** (2), 379 (1978).

20 Stankoviansky, S., Beno, A., Toma, S., and Gono, E., *Chem. Zvesti*, **24**, 19 (1970); *Chem. Abstr.*, **74**, 49014z (1971).

21 Butkiewicz, K., *J. Electroanal. Chem. Interfacial Electrochem.*, **39**, 419 (1972).

22 Butkiewicz, K., *J. Electroanal. Chem. Interfacial Electrochem.*, **87** (1), 137 (1978).

23 Rusina, A., Volke, J., Cernak, J., Kovac, J., and Kollar, V., *J. Electroanal. Chem. Interfacial Electrochem.*, **50**, 351 (1974).

24 Nikulin, V. N., Kargina, N. M., and Kargin, Y. M., *Zh. Obshch. Khim.*, **44**, 2520 (1974).

25 Nikulin, V. N., Kargina, N. M., and Kargin, Y. M., *Zh. Obshch. Khim.*, **43**, 2712 (1973).

26 Butkiewicz, K., *J. Electroanal. Chem. Interfacial Electrochem.*, **39**, 407 (1972).

27 Prevost, C., Souchay, P., and Chauvelier, J., *Bull. Soc. Chim. France*, 714 (1951).

28 Tirouflet, J., and Corvaisier, A., *Bull. Soc. Chim. France*, 535 (1962).

29 Corvaisier, A., and Tirouflet, J., *Compt. Rend.*, **251**, 1641 (1960).

30 Lavrushin, V. F., Bezuglyi, V. D., Belous, G. G., and Tishchenko, V. G., *Zh. Obshch. Khim.*, **34**, 7 (1964).

31 Lavrushin, V. F., Bezuglyi, V. D., Belous, G. G., and Tishchenko, V. G., *Zh. Organ. Khim.*, **1**, 98 (1965); *Chem. Abstr.*, **62**, 14453g (1965).

32 Butkiewicz, K., *J. Electroanal. Chem. Interfacial Electrochem.*, **48**, 297 (1973).

PART THREE
Physical Properties

Chapter Eighteen

Properties of Chalcones and Their Heterocyclic Analogues

VISCOSITY

The specific viscosity of chalcone in benzene and carbon tetrachloride is reported to be higher compared to dihydrochalcone.[1] At higher temperature (ca. 60°) the decrease in specific viscosity is slightly less in benzene than in carbon tetrachloride.

RATE OF CRYSTALLIZATION

The rate of crystallization has been used as a criterion of the purity of chalcone.[2] The maximum rate of crystallization (MRC), expressed as mm/minute, has been reported[2] for the following chalcones:

Compound	m.p. (°C)	MRC (temp.)
Chalcone	54.7	5.21 (30°)
p'-Ethylchalcone	61.0	1.15 (25°)
p'-Propylchalcone	45.7	3.62 (25°)

ADSORPTION CHARACTERISTICS

Chalcone undergoes polarization and color change when these are adsorbed on surface-active materials, such as silica gel, acid-washed alumina, and Brockmann's alumina. These color changes along with the ease of their elution (by methanol) from these adsorbents[2] are described in the literature.[3]

A study of the distribution of 4-dimethylaminochalcone (DMC) in an artificial lipid membrane has been reported.[4] DMC molecules have been found to concentrate in the polar regions, near the surface of the lipid membrane.[4]

QUENCHING OF FLUORESCENCE

Hesperatin chalcone, in contrast to hesperidin or eriodictin, is able to quench the fluorescence of chrysene solution (in acetone), and this has been attributed to its ability to form a nonfluorescent complex.[5]

EUTECTIC AND MOLECULAR COMPOUND FORMATION

Chalcone forms a eutectic[6] with picric acid and with trichloroacetic acid. A detailed study is reported of the eutectic composition of the ternary system, viz., chalcone-β-naphthol-picric acid (cf. Table 1).[7]

Table 1 Eutectic Composition

Temp. (°C)	Chalcone	β-Naphthol	Picric Acid
70	51.5	43.0	5.5
84	58.5	16.0	25.5
103	7.5	7.5	8.5

The molecular compounds formed between chalcone (and 3,4-methylenedioxychalcone) and isomeric nitrophenols (and *o*-nitrotoluene) have been studied.[8,9]

MOLECULAR REFRACTION

Refractometric studies of chalcones and some of its derivatives are reported in the literature.[10,11]

POLYMORPHISM

The phenomenon of polymorphism is well known in the case of chalcones. The work reported on the subject is summarized in Table 2.

BASICITY

The basicity of chalcone has been determined in sulfolane and water. The pK_a value −5.55 is reported for chalcone in sulfolane solvent, compared to −5.73 in water.[6]

The basicity of chalcone in acetic acid is reported to be higher than in aqueous solution.[29] The elctron-donating substituents, hydroxy, methoxy, and methyl, increase the basicity, whereas the *para*-substituted halogens decrease it.[29]

The basicity of *trans*-chalcone in concentrated sulfuric acid has been shown to be higher than its corresponding *cis* isomer.[30]

Protolytic equilibrium in substituted chalcones[31,32] and thiophene analogues[33] have been studied in the acetic acid–sulfuric acid system, and the relative basicities of these chalcones have been compared.[31, 33]

Table 2 Polymorphic Forms of Chalcones

Compound	Melting Point of Polymorphic Forms (°C)[a]	Remarks	Ref.
Chalcone[b]	59, 57, 49, 48, 30,—	Several polymorphic forms are known; no difference is observed in the infrared spectrum of the poly-morphs, recorded in the range 2222–1220 cm⁻¹; dipole moment data (calcd.) for some polymorphic forms are given	12, 13, 15, 16, 22
_m_-Methylchalcone	68, 67, 66, 53		14
_m'_-Methylchalcone	61, 51		14
_p_-Methylchalcone	99, 96.5, 90	Exists in three polymor-phic forms	24, 27
_p'_-Methylchalcone	74, 5, 56.5, 55.5, 54.5, 45.5, 48, 44.5		26
_o_-Nitrochalcone	126, 123		14
_m_-Nitrochalcone	146, 145, 120		14
_m'_-Nitrochalcone	131, 110		14
α-Bromochalcone	—	Exhibits polymorphism	12
β-Hydroxychalcone	81, 78, 73	Exhibits polymorphism	12, 28
β-Methoxychalcone	81, 78, 65	X-ray powder diffraction, UV, and IR spectral data have been recorded	12, 17, 18
β-Ethoxychalcone	81, 78, 75, 63, 43	Polymorphism is due to the existence of rotational isomers	21, 25, 28
β-Propoxychalcone	75, 63, 59	X-ray powder diffraction and UV studies have been carried out	21
β-Methoxy-_p_-nitrochalcone	97, 90, 78–81, 66–9, 55–8	Polymorphism attributed to rotational isomers. UV data are reported	20
β-Methoxy-_p'_-nitrochalcone	121, 107, 104	Polymorphic forms (107, 104°) possess identical x-ray powder diffraction,	19 12, 13,

Table 2   *(Continued)*

Compound	Melting Point of Polymorphic Forms (°C)[a]	Remarks	Ref.
		while the polymorph, m.p. 121°, has a different pattern; results confirmed by UV and IR data	15, 16, 22
β-Methoxy-p'-methoxy-p-nitrochalcone	132, 87, 80		22
	60, 56		13

[a]A dash indicates that polymorphism has been observed, but the melting points of the polymorphs are not available.
[b]It is reported to be in a mesophase near its melting point[23]; it exhibits a rapid increase in transmission of the infrared radiation at this stage of phase change.

A correlation has been found between the protonation data and the σ constants.[31]

Substitution of the phenyl ring of chalcone by either furan[34] or thiophene moiety[35–37] increase the basicity. The effect is reported[34,37] to be pronounced when the heterocyclic moiety is located farther away from the carbonyl group. Comparative studies in basicities of heterocyclic chalcone analogues have been published.[38–40]

HYDROGEN BONDING

The energy of hydrogen bonding involved in phenol and various chalcones in carbon tetrachloride has been determined.[41] The thermodynamic data for the (1:1) association of the above system have been reported.[42] The association constants have been shown to be a sensitive measure of the proton-acceptor power of chalcones.[43]

Based on UV absorption data the energy of the hydrogen bond formed between various substituted chalcones and trichloroacetic acid has been found to lie in the range 1.79–2.40 kcal/mole.[44,45] Also the ΔH values,[46] calculated from the equilibrium constants at various temperatures, have been shown to vary 2–3 kcal/mole.

The hydrogen bonding of 2,3,4,6-tetrachlorophenol with various chalcones has been studied by IR and NMR spectroscopy.[43]

ELECTRONIC EFFECTS

The influence of several functional groups on the activity of the methyl group in chalcone has been investigated.[47]

The quantum-chemical interpretation of reactivity of substituted chalcones (benzene and ferrocene type) has been made.[48] The effect of substituents on the π-electron structure and reactivitiy of monosubstituted *trans*-chalcones has been studied.[49]

The influence of transmission of electronic effects in Z and E isomers of α-phenyl-4-substituted chalcones has been investigated.[50] Z and E isomers are reported to be insensitive to substituent effects.[50]

HALOCHROMISM

The phenomenon of halochromism is reported for substituted chalcones[51-56] and their heterocyclic analogues containing either a furan,[57,58] thiophene,[59,61] selenophene,[60] or quinoline[62] nucleus. These chalcones develop halochromic colors when wetted with concentrated sulfuric acid. The deepening in color may be rationalized in terms of the formation of carbonium ions in concentrated sulfuric acid.

$$C_6H_5-\overset{+OH}{\underset{\|}{C}}-CH=CH-C_6H_5 \longleftrightarrow C_6H_5-\underset{+}{\overset{OH}{\underset{|}{C}}}-CH=CH-C_6H_5 \longleftrightarrow etc.$$

Methoxychalcones,[56] for example, also exhibit halochromism with hydrochloric and phosphoric acids. The color is reported to deepen with increase in the number of methoxy groups.[56]

The influence of various substituents on the halochromy of these compounds in concentrated sulfuric acid is described in the literature.[51-54] In the case of the quinoline analogue of chalcone, the halochromic color deepens when the heterocyclic residue is located closer to the carbonyl group.[62]

REFERENCES

1   Studinger, H., Steinhofer, A., and Bauer, R. C., *Annalen*, **517**, 54 (1935).

2   Michel, J., *Bull. Soc. Chim. Belges*, **48**, 105 (1939); *Chem. Abstr.*, **33**, 7650[7] (1939).

3   Weitz, E., Schmidt, F., and Singer, J., *Z. Electrochem.*, **46**, 222 (1940).

4   Dobretsov, G. E., Petrov, V. A., Deev, A. I., and Vladimirov, Y. A., *Biofizika*, **20**, 1014 (1975); *Chem. Abstr.*, **84**, 55578j (1976).

5   McLaughlin, J. A., and Szent-Gyorgyi, A., *Enzymologia*, **16**, 384 (1954); *Chem. Abstr.*, **49**, 10066h (1955).

6   Campbell, N., and Woodham, A. A., *J. Chem. Soc.*, 843 (1952).

7   Asahina, T., and Yokoyama, K., *J. Chem. Soc. Japan*, **56**, 415 (1935).

8   Asahina, T., *J. Chem. Soc. Japan*, **54**, 527 (1933).

9   Asahina, T., *Bull. Chem. Soc. Japan*, **9**, 181 (1934).

10 Volovik, A. M., Tolmachev, V. N., and Lavrushin, V. F., *Visn. Kharkiv Univ.*, **73**, 85 (1971); *Chem. Abstr.*, **78**, 57241u (1973).

11 Tolmachev, V. N., Volovik, A. M., and Lavrushin, V. F., *Zh. Obshch. Khim.*, **44**, 1810 (1974).

12 Weygand, C., *Annalen*, **472**, 143 (1929).

13 Weygand, C., *Chem. Ber.*, **62B**, 2603 (1929).

14 Weygand, C., and Schächer, F., *Chem. Ber.*, **68B**, 227 (1935).

15 LeFevre, R. J. W., *J. Chem. Soc.*, 1037 (1937).

16 Guy, J., *Ann. Phys.*, (12), **4**, 704 (1949); *Chem. Abstr.*, **44**, 10439d (1950).

17 Eistert, B., Weygand, F., and Csendes, E., *Chem. Ber.*, **85**, 164 (1952).

18 Ikeda, K., *J. Chem. Soc. Japan*, **76**, 899 (1955).

19 Ikeda, K., *J. Chem. Soc. Japan*, **76**, 896 (1955).

20 Iimura, F., *Nippon Kagaku Zasshi*, **77**, 1851 (1956); *Chem. Abstr.*, **53**, 5192h (1959).

21 Ikeda, K., *Nippon Kagaku Zasshi*, **78**, 302 (1957); *Chem. Abstr.*, **53**, 5192d (1959).

22   Guy, J., *Bull. Soc. Chim. France*, 731 (1949).

23 Taschek, R. F., and Williams, D., *J. Chem. Phys.*, **7**, 11 (1939).

24 Weygand, C., and Matthes, A., *Chem. Ber.*, **59B**, 2247 (1926).

25 Weygand, C., and Hennig, H., *Chem. Ber.*, **59B**, 2249 (1926).

26 Weygand, C., and Baumgartel, H., *Annalen*, **469**, 225 (1929).

27 Weygand, C., *Chem. Ber.*, **60B**, 2428 (1927).

28 Weygand, C., Bauer, E., and Hennig, H., *Chem. Ber.*, **62B**, 562 (1929).

29 Tsukerman, S. V., Kutulya, L. A., Lavrushin, V. F., and Thuy, N. M., *Zh. Fiz. Khim.*, **42**, 1930 (1968); *Chem. Abstr.*, **70**, 19471w (1969).

30 Noyce, D. S., and Jorgenson, M. J., *J. Am. Chem. Soc.*, **84**, 4312 (1962).

31 Trusevish, N. D., Lavrushin, V. F., Tolmachev, V. N., and Borodin, V. N., *Izv. Vyssh. Ucheb. Zaved., Khim. Khim. Tekhnol.*, **16**, 1677 (1973); *Chem. Abstr.*, **80**, 59270t (1974).

32 Trusevich, N. D., Tolmachev. V. N., and Lavrushin, V. F., *Vestn. Kharkov unta, Khim.*, **127**, 98 (1975); *Chem. Abstr.*, **84**, 134982q (1976).

33 Konate, B., Trusevish, N. D., Nikitchenko, V. M., and Lavrushin, V. F., *Izv. Vyssh. Ucheb. Zaved., Khim. Khim. Tekhnol.*, **18**, 1703 (1975); *Chem. Abstr.*, **84**, 73484e (1976).

34 Tsukerman, S. V., Kutulya, L. A., and Lavrushin, V. F., *Khim. Geterotsikl Soedin., Akad. Nauk. Latv. SSR* (6), 803 (1965); *Chem. Abstr.*, **64**, 12521e (1966).

35 Tsukerman, S. V., Kutulya, L. A., Nikitchenko, V. M., and Lavrushin, V. F., *Tr. Konf. Po. Probl. Primeneniya Korrelyatsion Uravnenii v Organ. Khim. Tartu*, **1**, 309 (1962); *Chem. Abstr.*, **61**, 1740g (1964).

36 Tsukerman, S. V., Kutulya, L. A., Surov, Y. N., Lavrushin, V. F., and Yur'ev, Y. K., *Dokl. Akad. Nauk. SSR*, **164**, 354 (1965); *Chem. Abstr.*, **63**, 18006c (1965).

37 Tsukerman, S. V., Kutulya, L. A., and Nikitchenko, V. M., *Zh. Obshch. Khim.*, **33**, 3180 (1963).

38 Tsukerman, S. V., Kutulya, L. A., Surov, Y. N., and Lavrushin, V. F., *Khim. Geterotsikl. Soedin.* (9), 1176 (1970); *Chem. Abstr.*, **74**, 87024k (1971).

39 Tsukerman, S. V., Surov, Y. N., Lavrushin, V. F., and Yurév, Y. K., *Khim. Geterotsikl. Soedin.* (6), 868 (1966); *Chem. Abstr.*, **67**, 32232b (1967).

40 Tsukerman, S. V., Kutulya, L. A., Lavrushin, V. F., and Yur'ev, Y. K., *Khim. Geterotsikl. Soedin., Akad. Nauk. Latv. SSR* (3), 376 (1966); *Chem. Abstr.*, **65**, 12093h (1966).

41 Katsui, G., and Kuyama, H., *Vitamins* (*Kyoto*), **11**, 197 (1956); *Chem. Abstr.*, **51**, 18472i (1957).

42 Gramstad, T., *Spectrochim. Acta*, **19**, 497 (1963).

43 Tsukerman, S. V., Kutulya, L. A., Surov, Y. N., Pivnenko, N. S., and Lavrushin, V. F., *Zh. Obshch. Khim.*, **40**, 1337 (1970).

44 Lavrushin, V. F., Tolmachov, V. M., Sinyagovskaya, L. A., and Trusevich, N. D., *Zh. Obshch. Khim.*, **35**, 1534 (1965).

45 Lavrushin, V. F., Tolmachov, V. M., Trusevich, N. D., and Sinyagovskaya, L. A., *Zh. Obshch. Khim.*, **35**, 1730 (1965).

46 Lavrushin, V. F., Trusevich, N. D., Tolmachev, V. N., and Semenov, A. I., *Zh. Obshch. Khim.*, **39**, 42 (1969).

47 Villard, C., *Bull. Soc. Fribourg. Sci. Nat.*, **47**, 20 (1957); *Chem. Abstr.*, **54**, 14238d (1960).

48 Zahradnik, P., and Lĕska, J., *Coll. Czech. Chem. Commun.*, **36**, 44 (1971).

49 Lĕska, J., and Zahradnik, P., *Coll. Czech. Chem. Commun.*, **38**, 3365 (1973).

50 Duke, P. J., and Boykin, D. W., Jr., *J. Org. Chem.*, **37**, 1436 (1972).

51 Kauffmann, H., and Kieser, F., *Chem. Ber.*, **46**, 3788 (1913).

52 Dilthey, W., Neuhaus, L., Reis, E., and Schommer, W., *J. Prakt. Chem.*, **124**, 81 (1930).

53 Dilthey, W., Blankenburg, C., Bradt, W., Braun, W., Dinklage, R., Huthwelker, W., and Schommer, W., *J. Prakt. Chem.*, **129**, 189 (1931).

54 Pfeiffer, P., and Kleu, H., *Chem. Ber.*, **66B**, 1058 (1933).

55 Lavrushin, V. F., and Verkhovod, N. N., *Ukr. Khim. Zh.*, **40**, 506 (1974); *Chem. Abstr.*, **81**, 119418e (1974).

56 Kuroda, C., and Matsukuma, T., *Sci. Papers Inst. Phys. Chem. Research (Tokyo)*, **18**, 51 (1932); *Chem. Abstr.*, **26**, 2442^8 (1932).

57 Lavrushin, V. F., Tsukerman, S. V., and Artemenko, A. I., *Zh. Obshch. Khim.*, **32**, 2551 (1962).

58 Tsukerman, S. V., Artemenko, A. I., and Lavrushin, V. F., *Zh. Obshch. Khim.*, **33**, 3528 (1963).

59 Tsukerman, S. V., Nikitchenko, V. M., and Lavrushin, V. F., *Zh. Obshch. Khim.*, **33**, 1255 (1963).

60 Tsukerman, S. V., Orlov, V. D., and Lavrushin, V. F., *Khim. Str., Svoistva Reaktivnost Org. Soedin.*, 63 (1969); *Chem. Abstr.*, **72**, 99698e (1970).

61 Lavrushin, V. F., Pogonina, R. I., and Izvekov, V. P., *Khim. Geterotsikl. Soedin.*, **7**, 1361 (1971); *Chem. Abstr.*, **76**, 58463p (1972).

62 Tsukerman, S. V., Ch'ang, K. S., and Lavrushin, V. F., *Zh. Obshch. Khim.*, **35**, 1723 (1965).

Chapter Nineteen

Complexing Action of Chalcones and Their Derivatives

Chalcones serve as starting materials for the preparation of ligands suitable for the quantitative precipitation of several metallic ions. This therefore forms the basis for their gravimetric estimation. The following two examples[1-4] are illustrative.

2'-Hydroxy-4-methoxy-5'-methylchalcone oxime[1-3] has been employed for the gravimetric estimation of divalent ions of palladium, copper, and nickel either separately or in combination.

2,4,6-Triphenylpyrylium chloride,[4] preparable from chalcone, has been used as a precipitant in the gravimetric estimation of gold and platinum.

A method of separation of palladium(II) from copper(II), involving the use of 2'-hydroxy-5'-methylchalcone oxime (as complexing agent), has been developed.[5] It is based on extraction, with a suitable solvent, of the resulting organometallic complex at a definite pH. The constituents thus separated have been determined spectrophotometrically.[5] 2'-Hydroxy-3'-bromo-4-methoxy-5'-methylchalcone oxime has been successfully employed in place of the aforesaid complexing agent.[6]

A spectrophotometric method of estimating beryllium, using 2'-hydroxychalcone, has been described.[7] Hydroxychalcones have been exploited as specific precipitants[8] (in the presence of EDTA) for

158

beryllium when present in combination with aluminum. Copper is reported to react with 2'-hydroxychalcone to form a complex that has a 1:2 stoichiometry.[9] On the basis of this reaction a conductometric method of estimation of copper has been developed.[9] Calcium has been determined by complexometric titration by using chalcone–metanil as indicator.[10]

The 1:1 complexes formed by the interaction of 2,2'-dihydroxy-chalcone (sodium salt) with the ions of copper(II), nickel(II), and tin(IV) have been studied.[11-14] The formation constants of several 1:1 complexes derived from the interaction of several metallic ions with 2'-hydroxychalcone[15,16] and 2',4'-dihydroxychalcone[17] have been determined.

The preparation of complexes derived from the reaction of 2'-hydroxychalcones with tri- and tetracarbonyl of iron has been reported.[18]

There are several preparation, spectroscopic, and conductometric studies[19,22-24,26-31] of the intermolecular complexes formed between chalcones and the halides of the following elements: Ti,[19-21] Fe,[26] Zn,[24] La,[23]Ce,[21] Pr,[23] Sm,[23] Eu,[23] B,[24,25] Al,[26] Si,[22] Ge,[22] Sn,[22,26] and Sb.[26-31]

The complexes of chalcones with antimony pentachloride have, however, been studied in detail. On the basis of conductometric measurements the existence of the following species has been proposed[27]:

$$n\text{SbCl}_5 + \text{Chalcone} \rightleftharpoons \text{Chalcone (SbCl}_5)_n \rightleftharpoons$$
$$\left[\text{Chalcone (SbCl}_4)_n\right]^{+1} \text{Cl}_n^{-1} \rightleftharpoons \left[\text{Chalcone (SbCl}_4)_n\right]^{+1} + n\text{Cl}^{-1}$$

where $n = 1$ or 2.

Two factors of paramount importance in the conductivity of the above complex[27] in benzene solution are temperature and the electron-donating ability of the substituents in the chalcone component.

REFERENCES

1 Deshmukh, B. K., Vyas, C. N., and Kharat, R. B., *J. Indian Chem. Soc.*, **52**, 385 (1975).

2 Deshmukh, B. K., Gholse, S. B., and Kharat, R. B., *Fresenius Z. Anal. Chem.*, **279**, 363 (1976).

3   Deshmukh, B. K., and Kharat, R. B., *Indian J. Chem.*, **14A**, 214 (1976).

4   Chadwick, T. C., *Anal. Chem.*, **46**, 1326 (1974).

5   Deshmukh, B. K., and Kharat, R. B., *Fresenius Z. Anal. Chem.*, **276**, 299 (1975).

6   Deshmukh, B. K., and Kharat, R. B., *J. Indian Chem. Soc.*, **56**, 213 (1979).

7   Naidu, R. R., *Proc. Indian Acad. Sci.*, **75A**, 124 (1972).

8   Naidu, R. R., *Talanta*, **22**, 614 (1975).

9   Naidu, R. S., and Naidu, R. R., *Proc. Indian Acad. Sci.*, **82A**, 142 (1975).

10   Yavorskaya, G. M., Kazak, R. V., and Lebedev, O. P., *Otkrytiya, Izobret. Prom. Obraztsy. Tovarnye. Znaki.*, **55**, 147 (1978); *Chem. Abstr.*, **89**, 70457r (1978).

11   Biradar, N. S., Patil, B. R., and Kulkarni, V. H., *Monatsh. Chem.*, **107**, 251 (1976).

12   Biradar, N. S., Patil, B. R., and Kulkarni, V. H., *Inorg. Chim. Acta*, **15**, 33 (1975).

13   Biradar, N. S., Patil, B. R., and Kulkarni, V. H., *Curr. Sci.*, **45**, 203 (1976).

14   Biradar, N. S., Patil, B. R., and Kulkarni, V. H., *Rev. Roum. Chim.*, **22**, 1479 (1977); *Chem. Abstr.*, **88**, 98463t (1978).

15   Khadikar, P. V., Kakkar, S. N., and Berge, D. D., *Indian J. Chem.*, **13**, 844 (1975).

16   Kakkar, S. N., and Khadikar, P. V., *J. Indian Chem. Soc.*, **54**, 667 (1977).

17   Khadikar, P. V., Kakkar, S. N., and Berge, D. D., *Indian J. Chem.*, **12**, 1319 (1974).

18   Brodie, A. M., Johnson, B. F. G., Josty, P. L., and Lewis, J., *J. Chem. Soc. Dalton Trans.* (18), 2031 (1972).

19   Evard, F., *Compt. Rend.*, **196**, 2007 (1933).

20   Biradar, N. S., and Patil, B. R., *Rev. Roun. Chim.*, **23**, 1425 (1978); *Chem. Abstr.*, **90**, 114211n (1979).

21   Biradar, N. S., and Angdi, S. D., *Monatsh. Chem.*, **109**, 1365 (1978).

22   Shelepina, V. L., Osipov, O. A., and Shelepin, O. E., *Issled. Termogr. Korroz.*, 201 (1970); *Chem. Abstr.*, **76**, 20792g (1972).

23   Spacu, P., and Plostinaru, S., *Rev. Roum. Chim.*, **14**, 591 (1969); *Chem. Abstr.*, **71**, 105874g (1969).

24   Efimov, A. A., Betin, O. I., Panasenko, A. I., and Starkov, S. P., *Zh. Prikl. Spektrosk.*, **23**, 1054 (1975); *Chem. Abstr.*, **84**, 97290q (1976).

25   Surov, Y. N., Shkumat, A. P., Nikitchenko, V. M., Tsukerman, S. V., and Lavrushin, V. F., *Zh. Obshch. Khim.*, **49**, 873 (1979).

26   Tomas, F., Carpena, O., and Mataix, J., *An. Quim.*, **68**, 115 (1972); *Chem. Abstr.*, **77**, 75418r (1972).

27   Tolmachev, V. N., Volovik, A. M., and Lavrushin, V. F., *Zh. Obshch. Khim.*, **40**, 275 (1970).

28   Yushko, V. K., Tolmachev, V. N., and Lavrushin, V. F., *Zh. Obshch. Khim.*, **40**, 1518 (1970).

29 Tolmachev, V. N., Volovik, A. M., and Lavrushin, V. F., *Zh. Neorg. Khim.*, **14**, 1572 (1969); *Chem. Abstr.*, **71**, 64891q (1969).

30 Lavrushin, V. F., Yushko, V. K., and Tolmachev, V. N., *Zh. Obshch. Khim.*, **40**, 156 (1970).

31 Yushko, V. K., Tolmachev, V. N., and Lavrushin, V. F., *Zh. Obshch. Khim.*, **40**, 160 (1970).

Chapter Twenty

Color Reactions, Detection, and Estimation of Chalcones

COLOR REACTIONS

Sulfuric Acid–Nitric Acid[1]

When the intensely colored solution of chalcone in concentrated sulfuric acid is treated with a little concentrated nitric acid, characteristic color changes occur. This is typified by the following examples:

Chalcone:	Orange $\longrightarrow$ yellow
4-Methoxychalcone:	Orange red $\longrightarrow$ yellow
Cinnamylideneacetophenone:	Cherry red $\longrightarrow$ dark yellow

This change involves nitration of chalcone rather than oxidation, and the resulting nitrochalcones exhibit weaker halochromy, as compared to unsubstituted chalcone.

Sulfuric Acid–Acetic Anhydride

The differing halochromic effects produced with concentrated sulfuric acid serve to characterize chalcones. Carbonium ions are believed to be formed as intermediates in this reaction. A large bathochromic shift in the visible color appears if acetic anhydride is incorporated with sulfuric acid. On treatment with a 200:1 (v/v) mixture of acetic anhydride–sulfuric acid, chalcones range from orange to purple[2]:

3,4,4'-Trihydroxychalcone	Orange
3,4'-Dihydroxy-4-methoxychalcone	Cerise
3,4,4'-Tribenzyloxychalcone	Red
3,4,4'-Trimethoxychalcone	Purple

The bathochromic shift arising due to the addition of acetic anhydride to the chalcones (in concentrated sulfuric acid) has been rationalized in terms of stability conferred on the carbonium ion by acetylation with acetic anhydride.[3] This is illustrated with reference to 4,4'-dimethoxychalcone.

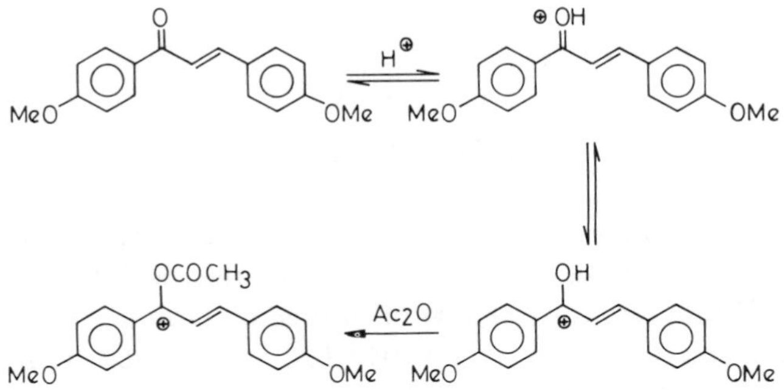

Sodium Borohydride and Hydrochloric Acid

Transient colors are developed when chalcones, after reduction with sodium borohydride, are treated with concentrated hydrochloric acid.

This color test can therefore be utilized for their identification. Table 1 lists the absorption maxima recorded àfter the chalcones were subjected to the aforesaid test[3]:

Table 1 Absorption Maxima of Chalcones Treated with NaBH$_4$ and HCl

Chalcone	λ_{max}(nm)
4'-Hydroxychalcone	480–485
4-Methoxychalcone	550
4,4'-Dimethoxychalcone	565
2'-Hydroxy-4,4',6-trimethoxychalcone	540
2'-Hydroxy-3,4,4',6'-tetramethoxychalcone	560
4-Dimethylaminochalcone	495
2'-Hydroxy-4,4-dimethylaminochalcone	550

WILSON'S BORIC ACID TEST

In the Wilson test a coloration appears in chalcones reacting with a boric acid–citric acid mixture in acetone solution. Partially methylated hesperetin chalcone is reported to give a positive test.[4] The boric acid–citric acid test is very specific for 5-hydroxy- and methoxychalcones.[5]

ANTIMONY PENTACHLORIDE TEST[6]

Various chalcones treated with antimony pentachloride in carbon tetrachloride yield intense red or violet precipitates, which are characteristically different from the yellow or orange precipitates from flavones, flavanones, and flavonols. A positive test is obtained with the chalcones listed in Table 2.[6]

The antimony pentachloride reaction is an extremely sensitive test. For example, 2-hydroxy-3,4,4'-trimethoxychalcone can be detected in a concentration as low as 1 ppm.

The color reactions of chalcones with various reagents are reported in the literature.[7]

Table 2   Colors Obtained with Various Chalcones in the SbCl$_5$ Test

Chalcone	Color
2-Hydroxy-2-methoxychalcone	Blood red
2-Hydroxy-4,5-dimethoxychalcone	Red
2-Hydroxy-4,4'-dimethoxychalcone	Red
2-Hydroxy-4',4,5-trimethoxychalcone	Blood red
2-Hydroxy-3,4,4'-trimethoxychalcone	Red
2-Hydroxy-3,4,5-trimethoxychalcone	Red
2-Hydroxy-4,5-dimethoxy-3',4'-methylenedioxychalcone	Cherry red
2-Hydroxy-3,4,4',6-tetramethoxychalcone	Red
2-Hydroxy-3,3',4,4',5-pentamethoxychalcone	Cherry red
2-Hydroxy-3,4,4',5-tetramethoxychalcone	Brick red
2,2'-Dihydroxy-3',4',5,6,6'-pentamethoxychalcone	Violet red
3,6-Dihydroxy-2,4,4'-trimethoxychalcone	Violet red
4-Methoxychalcone	Red
3',4'-Dimethoxychalcone	Cherry red
3',4'-Methylenedioxychalcone	Violet red
2',4,5-Trimethoxychalcone	Red
2,4,4',6-Tetramethoxychalcone	Dark red
2,3,4,4',5-Pentamethoxychalcone	Red
2,3,4-Trimethoxy-3',4'-methylenedioxychalcone	Cherry red

DETECTION AND ESTIMATION OF CHALCONE(S)

A spot test for chalcone has been developed, based on the pyrolytic oxidative cleavage of chalcone by lead dioxide.[8] Benzaldehyde, the product of reaction, is identified (identification limit: 50γ) by the appearance of yellow with thiobarbituric acid and phosphoric acid.

Chalcone may also be characterized by chromatography of its 2,4-dinitrophenylhydrazone[9] (R_f, 0.54) on paper, impregnated with diethyl ether–N,N-dimethylformamide–tetrahydrofuran (85 : 15 : 4; v/v).

A scheme is reported for the qualitative identification of two chalcones, for example, methylchalcone and tetramethoxyeriodictyolchalcone, following separation by paper chromatography.[10] The chalcone (identification limit: 5γ) on the chromatogram is revealed as a yellow fluorescent spot after spraying it with isonicotinic acid hydrazide reagent.

Methods for the separation and subsequent identification of chalcone in lemon oil are described in the literature.[11,12] According to one method, TLC has been employed for the purpose, using ethylacetate–hexane (3:2) as the eluotropic solvent.[12] Various chromogenic reagents are available for identification, which include concentrated hydrochloric acid, antimony trichloride (chloroform solution), 2,4-dinitrophenylhydrazine, and nicotinic acid hydrazide.

For quantitative estimation of chalcones in lemon oil the latter reacts with nicotinic acid hydrazide, and after 2 hours absorbance of the reaction mixture is recorded at $\sim$404 nm by means of a spectrophotometer.[12] Alternatively, chalcones may be determined gravimetrically by utilizing their reaction with 2,4-dinitrophenylhydrazine.[13] A spectrophotometric method for the determination of 2',4'-dihydroxychalcone and 7-hydroxyflavanone has been developed.[14] For this purpose the absorbance of the alcoholic solution of the mixture is measured at 320 nm.

p-Substituted benzohydrazides ($X—C_6H_4CONHNH_2$; $X = NO_2$, Br, or I) have been recommended[15–17] as reagents for the identification of chalcone. The melting points of the hydrazones are claimed to differ sufficiently to make identification possible, without the determination of mixed melting points[17]:

Compound	m.p.
Chalcone-p-nitrobenzohydrazone	232°
Chalcone-p-bromobenzohydrazone	170°
Chalcone-p-iodobenzohydrazone	182°

REFERENCES

1 Reddelien, G., *Chem. Ber.*, **45**, 2904 (1912).
2 King, H. G. C., and White, T., *J. Chem. Soc.*, 3539 (1961).
3 Krishnamurty, H. G., and Seshadri, T. R., *Curr. Sci.*, **34**, 681 (1965).
4 Anderson, J. R., O'Brien, K. G., and Reuter, F. H., *Anal. Chem. Acta*, **7**, 226 (1952).
5 Rangaswami, S., and Seshadri, T. R., *Proc. Indian Acad. Sci.*, **16A**, 129 (1942).
6 Marini-Bettòlo, G. B., and Ballio, A., *Gazz. Chim. Ital.*, **76**, 410 (1946).

7   Paris, R., and Cornilleau, J., *Ann. Pharm. Franc.*, **13**, 192 (1955); *Chem. Abstr.*, **50**, 4457h (1956).

8   Feigl, F., Ben-Dor, L., and Cohen, R., *Mikrochim. Acta* (6), 1181 (1964).

9   Barber, E. D., and Sawicki, E., *Anal. Chem.*, **40**, 984 (1968).

10   Hawker, C. D., Margraf, H. W., and Weichselbaum, T. E., *Anal. Chem.*, **32**, 122 (1960).

11   Stanley, W. L., *J. Assoc. Offic. Agr. Chemists*, **44**, 546 (1961); *Chem. Abstr.*, **55**, 22649h (1961).

12   D'Amore, G., and Sergi, G., *Rass. Chim.*, **18**, 8 (1966); *Chem. Abstr.*, **65**, 6994e (1966).

13   Davey, W., and Gwilt, J. R., *Chem. Ind. (London)*, 911 (1956).

14   Panasenko, A. I., and Starkov, S. P., *Izv. Vyssh. Ucheb. Zaved., Khim. Khim. Tekhnol.*, **19**, 664 (1976); *Chem. Abstr.*, **85**, 103504b (1976).

15   Chen, P., *J. Chinese Chem. Soc.*, **3**, 251 (1935); *Chem. Abstr.*, **29**, 7229[8] (1935).

16   Wang, S-M., Kao, Chen-Heng; Kao, Chung-Hsi, and Sah, P. P. T., *Sci. Rept. Natl. Tsinghau Univ.*, **A3**, 279 (1935); *Chem. Abstr.*, **30**, 2875[8] (1936).

17   Sah, P. P. T., and Hsu, C-L., *Rec. Trav. Chim.*, **59**, 349 (1940); *Chem. Abstr.*, **35**, 4363[1] (1941).

Chapter Twenty-One

Stereoisometrism in Chalcones and Their Derivatives

INTRODUCTION

The physical (color, crystallography, and transmutation) and chemical properties of the stereoisomers of chalcone have been examined.[1] Several forms exist, presumably due to polymorphism and/or stereoisomerism.[2] The introduction of methoxy or ethoxy groups in the β-position of the chalcone molecule exert different effects on isomerism. It is believed that the mass of the groups about the olefinic system, particularly their spatial effects, determine the number of isomers obtainable in a given case.[2]

The relationship between polymorphism and stereoisomerism of substituted chalcones has been discussed.[3]

The α-bromo-β-methoxychalcone[4] exists in two stereoisomeric forms. The most stable form has the highest melting point and is easier to isolate.[4] The stereoisomers have been interconverted into each other under appropriate experimental conditions.[6]

α-Bromo-β-methoxy- (and β-ethoxy-) chalcones also exist in several crystalline forms with different melting points.[5] However, their homologue, α-bromo-β-propoxychalcone, exists only in one form.

CIS–TRANS ISOMERISM

There are various reports about the existence of geometrical isomers of chalcones.[7-12] The well-known examples are chalcone,[7-9] α-phenylchalcone,[11,20] β-tolylchalcone,[10] and α-phenylnitrochalcone.[12]

The partial transformation of *trans*-chalcones into stereoisomeric pairs has been accomplished either by irradiation of their solution (in pentane or isooctane) in sunlight or by treatment with anhydrous aluminum chloride.[13,23] The percentage of *cis* and *trans* isomers formed on irradiation is reported to vary greatly from one chalcone to another.[14] The reverse conversion from *cis* to its stereoisomeric pair can be effected thermally[7] or by acid catalysis.[8]

The ultraviolet spectra of *cis–trans* isomers of chalcones have been studied.[15-17] Their spectra are quite different, with a large extinction coefficient for the more stable *trans* isomer.

In the acid-catalyzed isomerization of *cis*- to *trans*- chalcone either of the following mechanisms (A and B) are involved.

MECHANISM OF ACID-CATALYZED ISOMERIZATION

The isomerization of *cis* to *trans* is subject to smooth acid catalysis and proceeds to completion. The mechanism of this transformation is as follows:

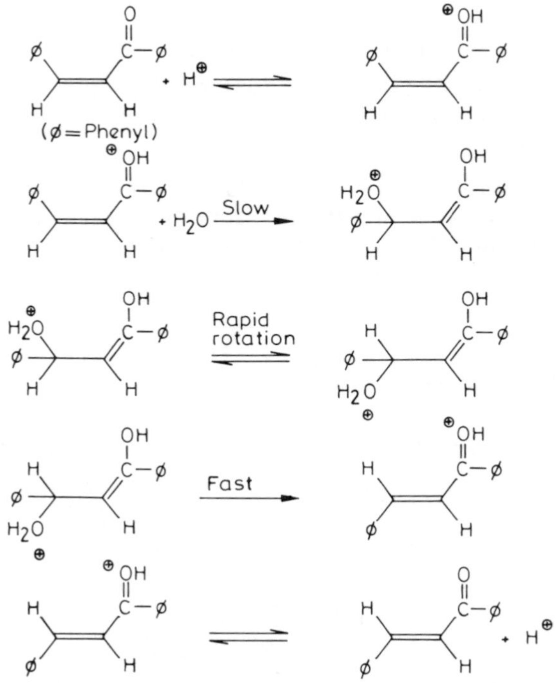

Mechanism A

The rate-controlling process is the addition of water to the conjugate acid (oxonium salt) of *cis*-chalcone. Following this slow step, rapid rotation about the C—C bond occurs, followed by dehydration. The mechanism outlined above finds support in the deuterium isotope studies on this transformation.[9]

Mechanism B (Carbonium Ion Mechanism)

A new mechanism is reported to intervene in the *cis–trans* isomerization of *para*-substituted chalcones (substituents: methoxy, nitro, or chloro) and is illustrated with respect to *cis*-4-methoxychalcone[18]:

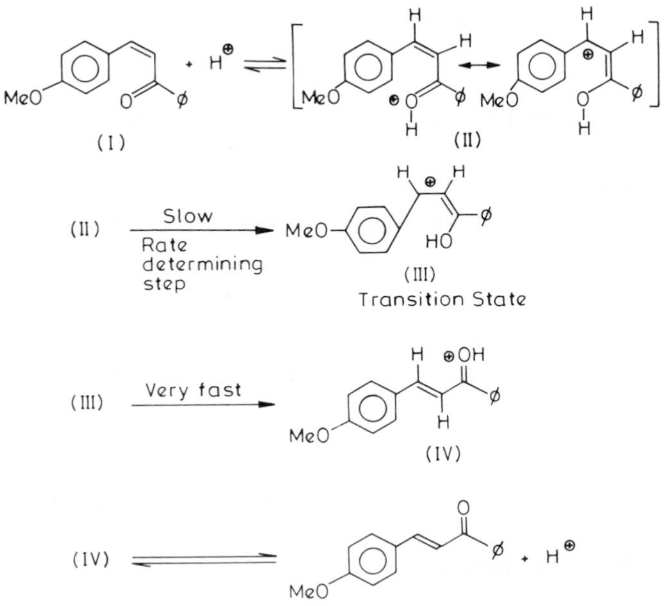

The salt of the ketone (**II**), is resonance-stabilized and undergoes rotation about C_α—C_β bond without the addition of water.

The isomerization may change from mechanism A to mechanism B, depending on the substituents present in the chalcone molecule, as well as on the conditions of the experiment.[19] The introduction of a methoxy group in the 2-, 4-, and 6-positions will exert a very great influence on the stability of the postulated benzylic carbonium ion (**II**) and thus favors isomerization by the latter mechanism. Also by reducing the activity of water the rate of isomerization by mechanism A may be reduced, thereby allowing mechanism B to become competitive. The introduction of *ortho* substituents will provide some steric interference on the addition of water at the β-carbon and likewise allow the carbonium ion mechanism to become competitive. The isomerization of *cis*-chalcone into *trans*-chalcone occurs in 78% sulfuric acid via mechanism A, while mechanism B operates when the reaction is carried out in a sufficiently concentrated sulfuric acid (96%).

The mechanism and kinetics of *cis–trans* photoisomerization of chalcone has been reported.[21] The reaction has been described to take place by a triplet mechanism, wherein a direct conversion of the twisted state of one isomer to the ground state of the other takes place.[21]

TAUTOMERISM IN α-HYDROXYCHALCONE

α-Hydroxychalcone is reported to exhibit tautomerism.[22] It exists in equilibrium with corresponding coumaran-3-one and the diketone,[22] thus

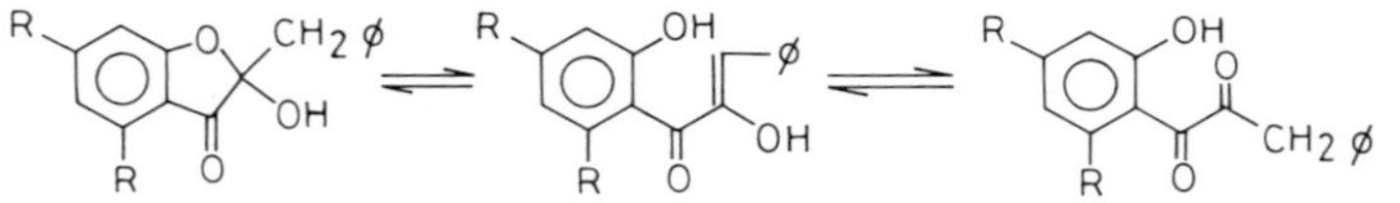

STEREOISOMERISM OF CHALCONE SEMICARBAZONES

Chalcone Semicarbazones

Chalcone reacts with semicarbazide in an acetic acid medium to give three isomeric semicarbazones (α, β, and γ).[24] According to the Hantzsch–Werner hypothesis four semicarbazones (**V–VIII**) are possible:

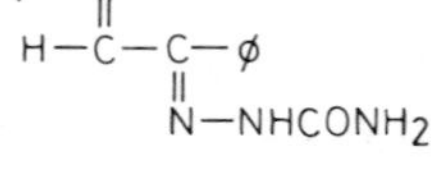

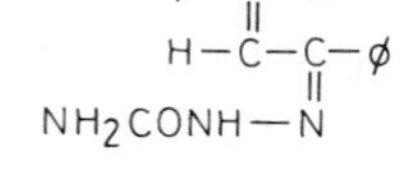

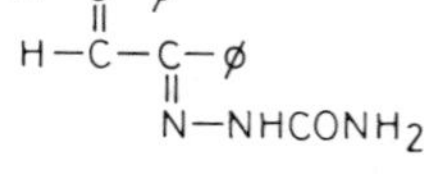

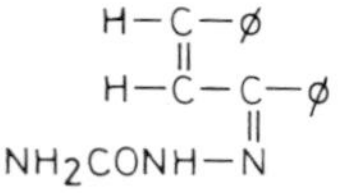

Under appropriate experimental conditions the mutual intercon-versions of α to γ form or γ to β form of semicarbazone are possible.[25] Similar behavior is exhibited by the isomers of chalcone phenylsemi-carbazone.[25]

Chalcone Thiosemicarbazones

Thiosemicarbazone of substituted chalcones (with substitution in both the aromatic rings) are described in the literature.[26] The thiosemicar-bazone of chalcone and 2-nitrochalcone are reported to exhibit proto-tropic and thermochromic properties.[27]

Chalcone semicarbazone and thiosemicarbazone formed under acid catalysis are reported to have *trans* and *anti* configurations with respect to the olefinic hydrogens.[28] On the other hand, *cis* and *syn* products are formed in the base-catalyzed reaction.

Chalcone semicarbazone derivatives are claimed to be useful for characterization of chalcone.[29,30] For the preparation of these deriva-tives the following two reagents have been recommended, 2,4-dinitro-phenyl semicarbazide[29] and α-naphthylsemicarbazide.[30] These reagents can readily be prepared by the interaction of hydrazine hydrate with the appropriately substituted urea in alcoholic solution.

REFERENCES

1 Dufraisse, C., *Ann. Chim. (France)*, **17**, 133 (1922); *Chem. Abstr.*, **16**, 2328[5] (1922).

2 Dufraisse, C., and Gillet, A., *Compt. Rend.*, **183**, 746 (1926).

3 Weygand, C., *Chem. Ber.*, **62B**, 562 (1929).

4 Dufraisse, C., and Gillet, A., *Compt. Rend.*, **178**, 948 (1924).

5 Dufraisse, C., and Gillet, A., *Ann. Chim. (France)*, **11**, 5 (1929); *Chem. Abstr.*, **23**, 4941[8] (1929).

6 Dufraisse, C., and Netter, R., *Compt. Rend.*, **189**, 299 (1929).

7 Lutz, R. E., and Jordan, R. H., *J. Am. Chem. Soc.*, **72**, 4090 (1950).

8 Noyce, D. S., Pryor, W. A., and King, P. A., *J. Am. Chem. Soc.*, **81**, 5423 (1959).

9 Noyce, D. S., Woo, G. L., and Jorgenson, M. J., *J. Am. Chem. Soc.*, **83**, 1160 (1961).

10 Badoche, M., *Bull. Soc. Chim. France*, **43**, 337 (1928).

11 Pascual-Vila, J., *An. Soc. Esp. Fis. Quim.*, **26**, 222 (1928); *Chem. Abstr.*, **23**, 381[9] (1929).

12   Buza, D., Gryff-Keller, A., and Szymanski, S., *Roczniki Chem.*, **45**, 549 (1971); *Chem. Abstr.*, **75**, 98071v (1971).

13   Dippy, J. F. J., McGhie, J. F., and Young, J. T., *Chem. Ind. (London)*, 195 (1952).

14   Hahn, L. V., and Miquel, J. F., *Compt. Rend.*, **257**, 1948 (1963).

15   Kuhn, L. P., Lutz, R. E., and Bauer, C. R., *J. Am. Chem. Soc.*, **72**, 5058 (1950).

16   Black, W. B., and Lutz, R. E., *J. Am. Chem. Soc.*, **75**, 5990 (1953).

17   Trummer, I., *Magy. Tud. Akad. Kozp. Fiz. Kut. Intez. Kozl.*, **4**, 481 (1956); **5**, 3, 130 (1957); *Chem. Abstr.*, **55**, 17207b (1961).

18   Noyce, D. S., and Jorgenson, M. J., *J. Am. Chem. Soc.*, **83**, 2525 (1961).

19   Noyce, D. S., and Jorgenson, M. J., *J. Am. Chem. Soc.*, **85**, 2420 (1963).

20   Friedrich, L. E., and Cormier, R. A., *J. Org. Chem.*, **35**, 450 (1970).

21   Mitina, V. G., Zadorozhnyi, B. A., and Lavrushin, V. F., *Zh. Obshch. Khim.*, **45**, 2713 (1975).

22   Enebäck, C., and Gripenberg, J., *Acta Chem. Scand.*, **11**, 866 (1957).

23   Davey, W., and Gwilt, J. R., *J. Chem. Soc.*, 1017 (1957).

24   Heilbron, I. M., and Wilson, F. J., *Proc. Chem. Soc.*, **27**, 315; *J. Chem. Soc.*, **101**, 1482 (1912).

25   Heilbron, I. M., and Wilson, F. J., *J. Chem. Soc.*, **103**, 1504 (1913).

26   Bu-Hoi, N. P., and Michel, S., *Bull. Soc. Chim. France*, 219 (1958).

27   Gheorghiu, C. V., and Avramovici, S., *An. Stiint. Univ. "Al. I. Cuza" Iasi, Sect. I [N.S.]*, **3**, 381 (1957); *Chem. Abstr.*, **53**, 15059c (1959).

28   Avramovici, S., Gabe, I., and Zugravescu, I., *Rev. Roumaine Chim.*, **10**, 471 (1965); *Studii Cercetari Chim.*, **14**, 453 (1965); *Chem. Abstr.*, **63**, 11317f (1965).

29   McVeigh, J. L., and Rose, J. D., *J. Chem. Soc.*, 713 (1945).

30   Sah, P. P. T., and Chiang, S-H., *J. Chinese Chem. Soc.*, **4**, 496 (1936); *Chem. Abstr.*, **31**, 4266^9 (1937).

Chapter Twenty-Two

Stereochemistry of Chalcones and Chalcone Analogues

INTRODUCTION

The conformational features of several chalcones have been arrived at by taking recourse to physical methods. The experimental methods for such investigations include, for example, x-ray crystallography,[1-8] dipole moment determination,[9-30] infrared spectroscopy,[27] and nuclear magnetic resonance spectroscopy. Only the former two methods are detailed in this chapter.

X-RAY CRYSTALLOGRAPHIC STUDIES ON UNSUBSTITUTED CHALCONE

Crystal and molecular structure studies based on x-ray crystallographic data are reported for unsubstituted chalcone.[1] It consists of two essentially planar units, which are linked by a nonplanar *cisoid* 1,3-enone bridge twisted about the C—C bond by 16.9°. The assignment of the *cisoid* conformation to chalcone finds additional support by the measurement of its molar Kerr constant.[2]

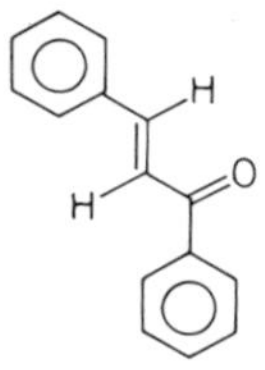

Polymorphic Forms of Chalcone

The conformational details about the polymorphic form of chalcone (m.p. 56°) have been worked out and compared with other polymorphic forms (m.p. 59°).[3]

4'-Bromochalcone

4'-Bromochalcone possesses a *cisoid* conformation, and there exists a large bond twist (26°) about $C_{(9)}$—$C_{(11)}$, and the two phenyl rings are situated at an angle of 50° with respect to each other.[4]

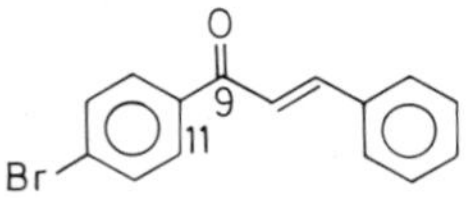

4-Methoxychalcone

X-ray crystallographic studies of 4-methoxychalcone indicate that the molecule consists of three planar units, methoxyphenyl, enone, and phenyl, the angle between the planar units being 4.5° and 11.5°.[5] A theoretical discussion has been published concerning the conformation of 4- and 4'-substituted chalcones.[6]

4,4'-Dimethylchalcone

X-ray studies on 4,4'-dimethylchalcone[7] reveal that the angle between the planes of the two phenyl rings is 48.6°, which represents a considerable deviation from planarity. This molecular conformation, it is claimed, controls the formation of optically active dibromides in the bromination of a single crystal of 4,4'-dimethylchalcone.[7]

Halonitrochalcones

transoid Conformations are reported to exist in the case of 3-chloro-, 3-bromo-, and 4-bromo-2'-nitrochalcones.[8]

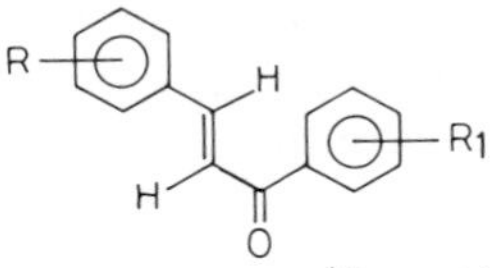

(Transoid)

DIPOLE MOMENT STUDIES ON CHALCONES

Dipole moments of a large number of chalcones[9-11] and their heterocyclic analogues[12-18] have been determined (cf. Table 1 below). These data have provided useful information regarding the stereochemistry of these molecules. From dipole moment measurement on chalcone, for example, it has been established that the double bond has a *trans* configuration, and the carbonyl group has a *S-cis* conformation with respect to the adjoining carbon–carbon double bond.[19]

Polymorphic Forms of Chalcone

It has been reported that the three polymorphic forms of chalcone (m.p. 59°, 57°, and 49°) have the same dipole moment.[20]

Bromochalcones

Dipole moment determination of 4'-bromochalcone, 4,4'-dibromochalcone, α,β-dibromochalcone, α-, and β-bromochalcones have been helpful in deciding whether the polar substituents are on the same or opposite sides of the double bond of the chalcone molecule.[9]

SUBSTITUTION EFFECTS ON DIPOLE MOMENT OF CHALCONES AND THEIR HETEROCYCLIC ANALOGUES

The effect of substituents on dipole moment of chalcone has been studied.[21] Based on the dipole moment data, the electron transfer through the carbonyl group of 4- and 4'-substituted hydroxychalcones has been determined.[22] With chalcone and 4-substituted chalcones (chloro-, bromo-, nitro-, and methoxy-) there is no interaction with the π-electron system of the phenyl ring (bearing the substituent) with the carbonyl group. In the case of 4,4-dimethylaminochalcone, however, there seems to be a direct polar conjugation between the carbonyl function and the dimethylamino group.

Enough data are available on the dipole moment of 4- and 4'-monosubstituted chalcone,[13] furan,[12,16,23,24] thiophene,[12,25] nitrothiophene,[26] selenophene,[12,15] pyrrole,[14] and pyridine[23] analogues of chalcone. The stereochemical implications of these data are summarized below.

Polarization caused by a heterocyclic ring is greater than that of phenyl, and the observed dipole moment is generally greater than the calculated value. Differences have been observed in 4- and 4'-monosubstituted chalcones and have been ascribed to the conjugation of the molecule.[13]

The following conclusions have been drawn in respect of pyrrole analogues of chalcone[14]:

1   All these compounds exist in *trans* form.

2   The carbonyl and the vinyl groups assume a *S-cis* conformation.

3   The *S-cis* conformation can exist in two rotational isomers, *syn* and *anti.* In the former case the NH of the pyrrole and carbonyl groups lie on the same side of the molecule, while in the latter case these lie on the opposite sides. The *syn* configuration predominates owing to the stability arising from the weak hydrogen bonding between the NH of pyrrole ring and the adjacent carbonyl group. In some of these compounds, however, the *anti* configuration is preferred, for example, 1-(2-pyrryl)-3-arylprop-3-ones and 1-(*N*-methyl-2-pyrryl)-3-phenyl-1-one. The *syn–anti* conformation is, however, preferred in the case of 1,3-di-(2-pyrryl)propenone.[14]

The selenophene analogues of chalcone of types I and II are reported to exist as mixtures of equal amounts of *syn-S-cis* and *anti-S-cis* conformers.[15]

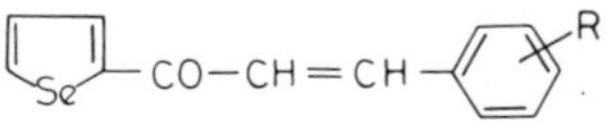

(I) 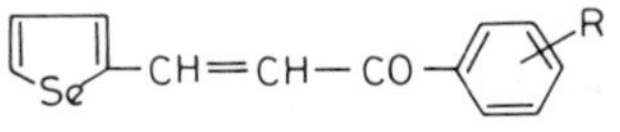   (II)

Similar work is reported for the substituted chalcones of the furan[16,23,24] series. The pyridine analogue, however, is reported to exist in a *transoid* conformation.[23]

Dipole moment data have been used to establish the configuration of the α-substituted α-(phenylthio)- and α-(phenylsulfonyl)chalcones.[10] Herein the benzoyl group is *cis* to the phenyl and *trans* to the hydrogen atom.

erythro-CHALCONE DIHALIDES

The infrared carbonyl absorption data for *erythro*-chalcone dihalides have been used for calculating the dipole moment of these compounds.[27]

CHALCONE EPOXIDES

The dipole moment data of chalcone epoxides are available in the literature.[28,29] These compounds are reported to exist in *gauche* conformation.[28] The oxides of the thiophene chalcone analogues exist in *S-cis* and *S-trans* conformations,[30] these differing in the mutual orientation of oxirane ring and carbonyl group.

HÜCKEL MOLECULAR ORBITAL CALCULATIONS

Based on the Hückel molecular orbital calculations the *trans* structure and *S-cis* stereochemistry of 2'-hydroxychalcones have been confirmed.[31]

Table 1 Dipole Moment of Chalcones

Compound	Dipole moment (D)	Ref.
Chalcone	2.92, 2.97, 3.03, 3.04	9, 13, 16, 32
4'-Chlorochalcone	2.98	13
4'-Bromochalcone	2.93	9
4'-Methoxychalcone	3.36	13
4'-Methylchalcone	3.19	13
4'-Nitrochalcone	4.21	13
4'-Phenylchalcone	3.12	13
4-Fluorochalcone	2.61	13
4-Chlorochalcone	2.54, 2.74	13, 32
4-Chloro-4'-methoxychalcone	3.62	33
4-Chloro-2',4'-dimethoxychalcone	4.09	33
4-Bromochalcone	2.47	9
4,4'-Dibromochalcone	2.03	9
α-Bromochalcone	3.87	9
β-Bromochalcone	3.59	9
4-Methoxychalcone	3.43, 3.40	13, 32
4,4'-Dimethoxychalcone	3.65	33
2',4,4'-Trimethoxychalcone	4.08	33
4'-Chloro-4-methoxychalcone	4.05	33
4'-Methyl-4-methoxychalcone	3.85	33
4'-Nitro-4-methoxychalcone	5.54	33
4'-Phenyl-4-methoxychalcone	3.83	33
4'-Chloro-2,4-dimethoxychalcone	4.85	33
2,4,4'-Trimethoxychalcone	4.31	33
4'-Methyl-2,4-dimethoxychalcone	4.29	33
4'-Nitro-2,4-dimethoxychalcone	7.04	33
4'-Phenyl-2,4-dimethoxychalcone	4.40	33
4-Methylchalcone	3.26	13
4'-Methoxy-4-methylchalcone	3.88	33
2',4'-Dimethoxy-4-methylchalcone	4.24	33
4-Nitrochalcone	3.6, 3.92	13, 32
4'-Methoxy-4-nitrochalcone	4.61	33
4,4-Dimethylaminochalcone	4.88	13
4-Phenylchalcone	3.03	13
4'-Methoxy-4-phenylchalcone	3.65	33
2',4'-Dimethoxy-4-phenylchalcone	4.32	33

Table 1 (Continued)

Compound	Dipole moment (D)	Ref.
$\gamma \quad \beta \quad \alpha$ X—CO—CH=CH—Y		
γ-Keto-γ-(2-furanyl)-α-phenyl-1-propene	3.19	16
γ-Keto-γ-(5-nitro-2-furanyl)-α-phenyl-1-propene	2.96	16
γ-Keto-γ-(2-furanyl)-α-(4-methoxyphenyl)-1-propene	3.59	16
γ-Keto-γ-(5-nitro-2-furanyl)-α-(4-methoxyphenyl)-1-propene	3.63	16
γ-Keto-γ-(2-furanyl)-α-(2,4-dimethoxyphenyl)-1-propene	4.18	16
γ-Keto--(5-nitro-2-furanyl)-(2,4-dimethoxyphenyl)α-1-propene	4.40	16
γ-Keto-γ-(2-furanyl)-α-(4-nitrophenyl)-1-propene	4.43	16
γ-Keto-γ-phenyl-α-(2-furanyl)-1-propene	3.03	16
γ-Keto-γ-(4-methoxyphenyl)-α-2-furanyl)-1-propene	3.22	16
γ-Keto-γ-(2,4-Dimethoxyphenyl)-α(2-furanyl)-1-propene	3.90	16
γ-Keto-γ-(4-nitrophenyl)-α-(2-furanyl)-1-propene	4.97	16
γ-Keto-γ-phenyl-α-(5-nitro-2-furanyl)-1-propene	2.91	16
γ-Keto-γ-(4-methoxyphenyl)-α-(5-nitro-2-furanyl)-1-propene	3.23	16
γ-Keto-γ-(2,4-dimethoxyphenyl)-α-(5-nitro-2-furanyl)-1-propene	5.17	16
γ-Keto-γ-(4-nitrophenyl)-α-(5-nitro-2-furanyl)-1-propene	4.86	16
γ-Keto-α, γ-di-(2-furanyl)-1-propene	3.17	16

REFERENCES

1   Rabinovich, D., *J. Chem. Soc., B*, 11 (1970).
2   Bramley, R., and Le Fevre, R. J. W., *J. Chem. Soc.*, 56 (1962).

3 Ohkura, K., Kashino, S., and Haisa, M., *Bull. Chem. Soc. Japan*, **46**, 627 (1973).

4 Rabinovich, D., Schmidt, G. M. J., and Shakked, Z., *J. Chem. Soc. Perkin Trans.*, **2**(1), 33 (1973).

5 Rabinovich, D., and Schmidt, G. M. J., *J. Chem. Soc.*, *B*, 6 (1970).

6 Tsukerman, S. V., Surov, Y. N., and Lavrushin, V. F., *Zh. Organ. Khim.*, **6**, 887 (1970); *Chem. Abstr.*, **73**, 14315a (1970).

7 Rabinovich, D., and Shakked, Z., *Acta Crystallogr.*, **30B**, 2829 (1974).

8 Jungk, A. E., and Schmidt, G. M. J., *J. Chem. Soc.*, *B*, 1427 (1970).

9 Bergmann, E., *J. Chem. Soc.*, 402 (1936).

10 Baliah, V., and Natarajan, C., *Indian J. Chem.*, **9**, 435 (1971).

11 Dinya, Z., Litkei, G., Levai, A., Boleskei, H., Jeckel, P., Rochlitz, S., Kiss, A. I., Farkas, M., and Bognár, R., *Bioflavonoid Symposium, 5th*, pp. 247–55 (1977). Edited by L. Farkas, M. Gabor, and F. Kalley, Elsevier, Amsterdam; *Chem. Abstr.*, **89**, 42239c (1978).

12 Tsukerman, S. V., Orlov, V. D., Nikitchenko, V. M., Rozum, Y. S., Lavrushin, V. F., and Yur'ev, Y. K., *Teor. Eksp. Khim., Akad. Nauk Ukr. SSR*, **2**, 399 (1966); *Chem. Abstr.*, **65**, 16839h (1966).

13 Tsukerman, S. V., Surov, Y. N., and Lavrushin, V. F., *Zh. Obshch. Khim.*, **38**, 524 (1968); *Chem. Abstr.*, **69**, 6536a (1968).

14 Tsukerman, S. V., Izvekov, V. P., and Lavrushin, V. F., *Zh. Fiz. Khim.*, **42**, 2159 (1968); *Chem. Abstr.*, **70**, 57025m (1969).

15 Tsukerman, S. V., Orlov, V. D., Lam Ngok Thiem, and Lavrushin, V. F., *Khim. Geterotsikl. Soedin.*, **6**, 974 (1969); *Chem. Abstr.*, **73**, 7447s (1970).

16 Tsukerman, S. V., Artemenko, A. I., and Lavrushin, V. F., *Zh. Obshch. Khim.*, **34**, 3591 (1964); *Chem. Abstr.*, **62**, 4736c (1965).

17 Tsukerman, S. V., Bugai, A. I., Izvekov, V. P., and Lavrushin, V. F., *Khim. Geterotsikl. Soedin.*, **8**, 1083 (1972); *Chem. Abstr.*, **77**, 151341c (1972).

18 Nikitchenko, V. M., Ibrahim, S., and Lavrushin, V. F., *Izv. Vyssh. Ucheb. Zaved., Khim. Khim. Tekhnol.*, **15**, 720 (1972); *Chem. Abstr.*, **77**, 87648d (1972).

19 Grunfest, M. G., Kolodyazhnyi, Y. V., Osipov, O. A., Yanovskaya, L. A., Umirzakov, B., Kucherov, V. F., and Vorontsova, L. G., *Izv. Akad. Nauk SSSR, Ser. Khim.* (12), 2662 (1972); *Chem. Abstr.*, **78**, 110356p (1973).

20 Eisenlohr, F., and Metzner, A., *Z. Physik. Chem.*, **178A**, 350 (1937); *Chem. Abstr.*, **31**, 4552^1 (1937).

21 Yanovskaya, L. A., Umirzakov, B., Kucherov, V. F., Yakovlev, I. P., Zolotarev, B. M., Chizhov, O. S., Vorontsova, L. G., Fundyler, I. N., Grunfest, M. G., Kolodyazhny, Y. V., and Osipov, O. A., *Tetrahedron*, **29**, 4321 (1973).

22 Verkhovod, N. N., Verkhovod, V. M., Litvinenko, E. V., and Lavrushin, V. F., *Visn. Kharkiv. Univ., Khim.*, **73**, 81 (1971); *Chem. Abstr.*, **78**, 57506j (1973).

23 Orlov, V. D., Tsukerman, S. V., Lavrushin, V. F., *Vop. Stereokhim. No. 1*, 89 (1971); *Chem. Abstr.*, **77**, 151343e (1972).

24 Andrieu, C., and Lumbroso, H., *C. R. Acad. Sci., Ser. C.*, (13), 272 (1971).

25 Lavrushin, V. F., Verkhovod, V. M., Roberman, A. I., and Ostrovskaya, B. I.,

Izv. Vyssh. Ucheb. Zaved., Khim. Khim. Tekhnol., **18**, 1707 (1975); *Chem. Abstr.*, **84**, 73485f (1976).

26 Verkhovod, V. M., Khashchina, M. V., Berestetskaya, V. D., and Lavrushin, V. F., *J. Structural Chem.*, **18**, 26 (1977).

27 Weber, F. G., and Westphal, G., *Z. Chem.*, **15**, 215 (1975).

28 Orlov, V. D., Korotkov, S. A., Tishchenko, V. N., and Lavrushin, V. F., *Zh. Strukt. Khim.*, **14**, 567 (1973); *Chem. Abstr.*, **79**, 65590t (1973).

29 Yanovskaya, L. A., Dombrovskii, V. A., Kucherov, V. F., Grunfest, M. G., Kolodyazhnyi, Y. V., and Osipov, O. A., *Izv. Akad. Nauk SSSR, Ser. Khim.* (9), 2144 (1973); *Chem. Abstr.*, **80**, 59277a (1974).

30 Arbuzov, B. A., Vul'fson, S. G., Donskova, A. I., and Vereshchagin, A. N., *Bull. Acad. Sci. USSR*, **25**, 1443 (1977).

31 Grouiller, A., Thomassery, P., Pacheco, H., Decoret, C., and Tinland, B., *Bull. Soc. Chim. France*, **12**, 3454 (1973).

32 Mauret, P., Mermillod-Blardet, D., and Maroni, P., *Bull. Soc. Chim. France* (3), 903 (1971).

33 Lavrushin, V. F., Verkhovod, N. N., Verkhovod, V. M., and Litvinenko, E. V., *Zh. Obshch. Khim.*, **40**, 1343 (1970).

Chapter Twenty-Three

Spectroscopic Studies of Chalcones and Their Derivatives

ULTRAVIOLET SPECTROSCOPY

Introduction

trans-Chalcone usually shows two absorption bands[1] located at 300 nm (band I) and 230 nm (band II). Previously these bands were believed to originate due to the presence of two chromophoric groups,[2] benzoyl and cinnamoyl moieties, in the chalcone molecule. According to the accepted view,[3] band I results from the conjugation of the whole molecule, thus

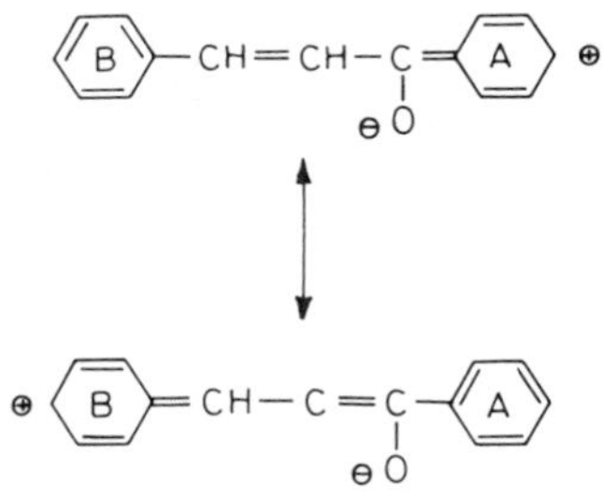

A third band at 205 nm has also been reported and has been characterized as the modified E band of the phenyl group.[4] Another band located at 250–270 nm (middle band) has been observed in *cis* chalcones,[2,3,5–8] as well as in some *trans* isomers.[1] It has been reported that the UV spectrum of *cis*-chalcone contains styrene absorption bands, which are absent in the spectrum of *trans*-chalcone.[9]

The electronic absorption spectra have been calculated for chalcone,[10] its derivatives,[11] and the heterocyclic analogues of chalcone.[12] On the basis of theoretical calculations it has been shown that the long-wave ultraviolet absorption bands of 4-substituted chalcones are more complex in nature.[12] There are other bands besides those arising due to intramolecular charge transfer transition.

Substituent Effects on UV Absorption of Chalcones[4,14–17]

In general, substitution in ring B causes a greater bathochromic shift of band I than in ring A.[4,14,15] Electron-withdrawing substituents in ring B produce large hypsochromic effect[16] but when present in ring A they exhibit bathochromic effects.

Replacement of the phenyl moiety (ring A or ring B) with 2-thienyl or 2-furyl gives rise to a bathochromic shift of band I.[3,18,19] The effect, however, is more pronounced when the phenyl group (ring B) is substituted by the heterocyclic residue.[19]

Solvent Effect

The effect of solvent on the position of $n-\pi^*$ and $\pi-\pi^*$ transition in chalcone derivatives has been studied.[20]

Miscellaneous Studies

The UV absorption studies are reported for the following compounds: chalcones,[1-3,8,10,13,15,21-57] heterocyclic analogues of chalcone,[16,19,58-69] chalcone glucosides,[70] chalcone complexes with boron trifluoride,[71] ferric chloride,[37,72] antimony pentachloride[72] and boric acid,[41] chalcone-2,4-dinitrophenylhydrazones,[73-76] and chalcols.[77]

INFRARED SPECTROSCOPY

Several studies on IR absorption of chalcones,[78-86] heterocyclic analogues of chalcones,[9,87-91] and chalcone derivatives[92] have been published.

Carbonyl Group, Substituent Effects

The integral intensities of the carbonyl absorption band in the infrared spectra of chalcone, 4- and 4'-substituted chalcones, and heterocyclic analogues (selenienyl, thienyl, and furyl) have been reported.[93]
    The frequency of carbonyl absorption in chalcone has been determined by Hückel's molecular orbital method as well as on the basis of other theoretical calculations.[94]
    The substitutent effects on the carbonyl stretching frequency of chalcone[95-98] and its heterocyclic analogues[90,93] have been published.

Hydrogen Bonding

The effect of intermolecular and intramolecular hydrogen bonding on the carbonyl group of hydroxychalcones has been described in the literature.[96,99,100]

Intramolecular hydrogen bonding in *cis* and *trans* isomers of 2'-hydroxy-3'-nitrochalcone has been studied.[101] In the *cis* isomer the hydroxyl and the nitro groups are hydrogen-bonded, while in the *trans* isomer the 2'-hydroxyl group is chelated with the carbonyl group.[101]

In the case of 2'-hydroxychalcone the hydrogen bond energy has been calculated.[95]

Basicity

Infrared absorption data have been used for comparison of relative basicities of chalcones,[102-104] ferrocene,[105] and thiophene[106] analogues of chalcone.

Rotational Isomers

Infrared spectroscopy has been employed in studying the rotational isomers (*S-cis* and *S-trans*) of substituted chalcones[107-117] and heterocyclic analogues of chalcone.[105,118]

Raman Spectroscopy

The Raman spectrum of chalcone has been reported in the literature.[119-121]

NUCLEAR MAGNETIC RESONANCE SPECTROSCOPY

Chalcones and Their Heterocyclic Analogues

The proton magnetic resonance spectra of some chalcones[122] and their heterocyclic analogues (containing furan, thiophene, and selenophene ring attached at 2 position) are reported in the literature.[123]

The correlation between the NMR signals of the hydroxyl groups of chalcone derivatives (in acetone solvent) and the σ substituent parameters have been reported.[124]

^{13}C-NMR chemical shifts of substituted chalcones[125] as well as those of heterocyclic analogues of chalcone[126] are available in the literature. ^{13}C-NMR and -PMR spectra have been studied for ferrocenyl chalcone and their iron carbonyl complexes.[127] It has been stated that the conjugation in the chalcone molecule is reduced by its coordination with the iron carbonyl. Based on ^{13}C-NMR data the effect of the ferrocenyl group on the π-electron distribution in chalcone has been studied.[128]

Chalcone Derivatives

From the examination of NMR spectra of some substituted chalcone epoxides the following conclusions have been drawn.[129]

The epoxide ring has a *trans* configuration, and the two oxirane ring protons appear in the range 5.70–5.82τ and 5.82–6.02τ, respectively.

NMR data are available for isocyanatohalochalcones, *trans*-chalconeoxime, semicarbazone, and thiosemicarbazone 3,5-diphenyl-1-acetamido- (and thioacetamido-) 2-pyrazolines.[92,130]

MASS SPECTROMETRY

Chalcones

The mass spectrum of chalcone shows three favored ions, M$^+$, (M-1)$^+$, and (M-29)$^+$ where M$^+$ stands for the molecular ion. The following mechanism has been advanced to explain the fragmentation process[131]:

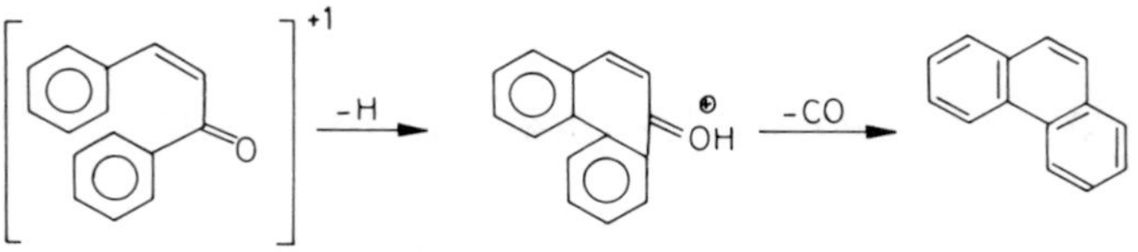

The first step is the loss of hydrogen, since it permits the formation of an extra bond. This is followed by the loss of carbon monoxide leading to the fused ring structure (see above).

Skeletal rearrangements occur in the mass spectrometry of chalcone. The following rearrangement ions are reported with their relative abundance (expressed as percent)[132]:

$$C_{13}H^+_9 (11) \qquad C_{14}H^+_{10} (13) \qquad C_{14}H^+_{11} (26)$$

It has been shown that electron bombardment of 2'-hydroxychalcone leads to the formation of flavanone.[133,134] The mechanism involves the migration of the phenolic hydrogen to the 3-position of the flava-none-type ion[133]:

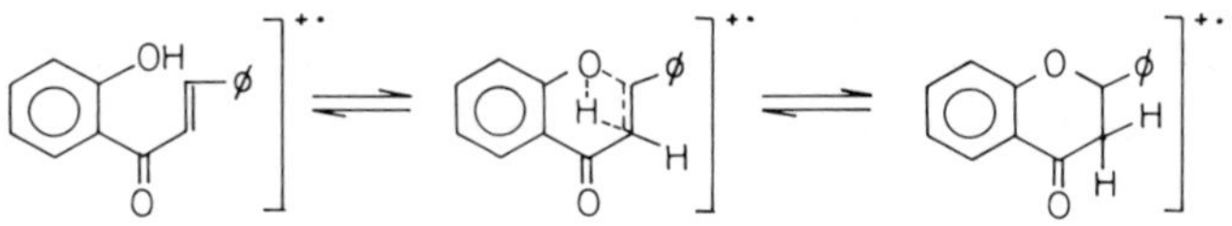

The main fragmentation of path of 2'-hydroxychalcone is illustrated here[135]:

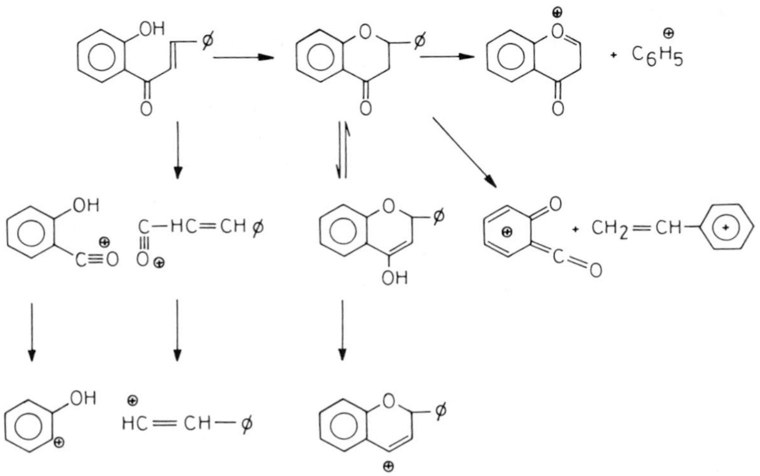

Based on mass spectrometric data the ionization potential of 4- and 4'-monosubstituted chalcones have been determined. The values so obtained are reported to be in agreement with the values arrived at by molecular orbital calculations.[136]

Chalcone Epoxides

Mass spectrometry has been utilized for the qualitative characterization of the transmission effects through the oxirane ring of chalcone epoxide.[137]

REFERENCES

1   Wheeler, O. H., Gore, P. H., Santiago, M., and Baez, R., *Can. J. Chem.*, **42**, 2580 (1964).

2   Katzenellenbogen, E. R., and Branch, G. E. K., *J. Am. Chem. Soc.*, **69**, 1615 (1947).

3   Szamant, H. H., and Basso, A. J., *J. Am. Chem. Soc.*, **74**, 4397 (1952).

4   Kline, P., and Gibian, H., *Chem. Ber.*, **94**, 26 (1961).

5   Lutz, R. E., and Jordan, R. H., *J. Am. Chem. Soc.*, **72**, 4090 (1950).

6   Lutz, R. E., Hinkley, D. F., and Jordan, R. H., *J. Am. Chem. Soc.*, **73**, 4647 (1951).

7   Black, W. B., and Lutz, R. E., *J. Am. Chem. Soc.*, **75**, 5990 (1953).

8   Black, W. B., and Lutz, R. E., *J. Am. Chem. Soc.*, **77**, 5134 (1955).

9   Mitina, V. G., Sukhorukov, A. A., Zadorozhnyi, B. A., and Lavrushin, V. F., *Zh. Obshch. Khim.*, **46**, 699 (1976).

10   Sukhorukov, A. A., Zadorozhnyi, B. A., and Lavrushin, V. F., *Teor, Eksp. Khim.*, **6**, 602 (1970); *Chem. Abstr.*, **74**, 93047m (1971).

11   Leska, J., and Zahradnik, P., *Acta Fac. Rerum Nat. Univ. Comenianae, Chim.*, No. 18, 43 (1973); *Chem. Abstr.*, **84**, 73422h (1976).

12   Reinhardt, M., Sukhorukov, A. A., Raushal, A., and Lavrushin, V. F., *Khim. Geterotsikl. Soedin.* (9), 1202 (1977); *Chem. Abstr.*, **88**, 5753z (1978).

13   Fundyler, I. N., and Ryaboi, V. M., *Izv. Akad. Nauk SSSR, Ser. Khim.* (3), 573 (1974); *Chem. Abstr.*, **81**, 43343d (1974).

14   Shibata, Y. A., and Nagai, W., *Acta Phytochim.*, **2**, 25; *Chem. Abstr.*, **19**, 3064[4] (1925).

15   Shibata, Y. A., and Nagai, W., *J. Chem. Soc. Japan*, **43**, 101 (1922); *Chem. Abstr.*, **16**, 2512[9] (1922).

16   Coleman, L. E., *J. Org. Chem.*, **21**, 1193 (1956).

17   David, F. R., Szabo, G. B., Rakosi, M., and Litkei, G., *Acta Chim. Acad. Sci. Hung.*, **94**, 57 (1977).

18   Szmant, H. H., and Basso, A. J., *J. Am. Chem. Soc.*, **73**, 4521 (1951).

19   Szmant, H. H., and Planinsek, H. J., *J. Am. Chem. Soc.*, **76**, 1193 (1954).

20   Verkhovod, V. M., Verkhovod, N. N., Tolmachev, V. N., and Lavrushin, V. F., *Visn. Kharkiv. Univ. Khim.*, **73**, 77 (1971); *Chem. Abstr.*, **78**, 57244x (1973).

21   Stobbe, H., and Ebert, E., *Chem. Ber.*, **44**, 1289 (1911).

22   Russell, A., Todd, J., and Wilson, C. L., *J. Chem. Soc.*, 1940 (1934).

23   Alexa, V., *Bul. Soc. Chim. Romania*, **18A**, 93 (1936); *Chem. Abstr.*, **31**, 3387[4] (1937).

24   Cromwell, N. H., and Johnson, R. S., *J. Am. Chem. Soc.*, **65**, 316 (1943).

25   Ferguson, L. N., and Barnes, R. P., *J. Am. Chem. Soc.*, **70**, 3907 (1948).

26   Cromwell, N. H., and Watson, W. R., *J. Org. Chem.*, **14**, 411 (1949).

27   Thomas, J. F., and Branch, G., *J. Am. Chem. Soc.*, **75**, 4793 (1953).

28 Black, W. B., *Diss. Abstr.*, **14**, 1532 (1954).

29 Browne, C. L., and Lutz, R. E., *J. Org. Chem.*, **18**, 1638 (1953).

30 Wiley, R. H., Jarboe, C. H., Jr., and Ellert, H. G., *J. Am. Chem. Soc.*, **77**, 5102 (1955).

31 Grammaticakis, P., and Chauvelier, *J. Compt. Rend.*, **242**, 1189 (1956).

32 Walker, E. A., and Young, J. R., *J. Chem. Soc.*, 2041 (1957).

33 Hassner, A., and Cromwell, N. H., *J. Am. Chem. Soc.*, **80**, 893 (1958).

34 Haske, B. J., *Diss. Abstr.*, **20**, 1168 (1959).

35 House, H. O., and Ryerson, G. D., *J. Am. Chem. Soc.*, **83**, 979 (1961).

36 Hahn, L. V., and Miquel, J. F., *Compt. Rend.*, **257**, 1948 (1963).

37 Lavrushin, V. F., Dzyuba, V. P., and Tolmachov, V. M., *Zh. Obshch. Khim.*, **35**, 95 (1965).

38 Sinyagovskaya, L. A., Lavrushin, V. F., and Tolmachov, V. M., *Zh. Obshch. Khim.*, **35**, 1929 (1965).

39 Bondarenko, V. E., Dashevskii, M. M., Krasovitskii, B. M., and Pereyaslova, D. G., *Zh. Organ. Khim.*, **2**, 1060 (1966).

40 Mahanta, B. C., Nayak, P. L., and Rout, M. K., *J. Inst. Chem. (India)*, **42**, 49 (1970); *Chem. Abstr.*, **73**, 55239x (1970).

41 Hörhammer, L., and Hänsel, R., *Arch. Pharm.*, **288**, 315 (1955); *Chem. Abstr.*, **51**, 6619h (1957).

42 Wittman, H., Uragg, H., and Stark, H., *Monatsh. Chem.*, **97**, 896 (1966).

43 Akaboshi, S., and Kutsuma, T., *Yakugaku Zasshi*, **89**, 375 (1969); *Chem. Abstr.*, **71**, 29883s (1969).

44 Nikitina, A. N., Fedyunina, G. M., Umirzakov, B., Yanovskaya, L. A., and Kucherov, V. F., *Opt. Spektrosk.*, **34**, 289 (1973); *Chem. Abstr.*, **78**, 158500h (1973).

45 Alexa, V., *Bul. Chim., Soc. Chim. Romania [2]*, **1**, 77 (1939); *Chem. Abstr.*, **37**, 3668⁴ (1943).

46 Lin, C-H., Su, J-T., and Chen, F-C., *Tai-wan K'o Hsueh*, **17**, 29 (1963); *Chem. Abstr.*, **60**, 7896c (1964).

47 Ristic, S., and Baranac, J. M., *Proc. Colloq. Spectrosc. Int. 14th*, **3**, 1219 (1967); *Chem. Abstr.*, **73**, 93177h (1970).

48 Dzurilla, M., Kristian, P., and Gyoryova, K., *Chem. Zvesti.*, **24**, 207 (1970); *Chem. Abstr.*, **74**, 53196c (1971).

49 Dhar, D. N., and Singhal, D. V., *Spectrochim. Acta*, **26A**, 1171 (1970).

50 Patel, D. R., and Patel, S. R., *J. Indian Chem. Soc.*, **45**, 703 (1968).

51 Cavina, G., *Rend. Ist. Super. Sanita*, **16**, 154 (1953); *Chem. Abstr.*, **48**, 8651f (1954).

52 Polansky, O., *Monatsh. Chem.*, **88**, 91 (1957).

53 Baliah, V., and Shanmuganathan, S., *J. Phys. Chem.*, **62**, 255 (1958).

54 Orlov, V. D., Borovoi, I. A., and Lavrushin, V. F., *Zh. Obshch. Khim.*, **43**, 642 (1973).

55 Dhar, D. N., and Singh, R. K., *J. Indian Chem. Soc.*, **50**, 129 (1973).

56   Dhar, D. N., and Singh, R. K., *J. Indian Chem. Soc.*, **49**, 241 (1972).

57   Dolter, R. J., and Columba, C., *J. Am. Chem. Soc.*, **82**, 4153 (1960).

58   Lavrushin, V. F., Tsukerman, S. V., and Nikitchenko, V. M., *Zh. Obshch. Khim.*, **32**, 2677 (1962).

59   Lavrushin, V. F., Tsukerman, S. V., and Artemenko, A. I., *Zh. Obshch. Khim.*, **32**, 2551 (1962).

60   Tsukerman, S. V., Artemenko, A. I., and Lavrushin, V. F., *Zh. Obshch. Khim.*, **33**, 3528 (1963).

61   Egger, H., and Schloegl, K., *J. Organometal. Chem.*, **2**, 398 (1964).

62   Trusevich, N. D., Kanate, B., Nikitchenko, V. M., and Lavrushin, V. F., *Dopov. Akad. Nauk. Ukr. RSR,. Ser. B*, **36**, 454 (1974); *Chem. Abstr.*, **81**, 62911d (1974).

63   Mitina, V. G., Zadorozhnyi, B. A., and Lavrushin, V. F., *Teor. Eksp. Khim.*, **9**, 263 (1973); *Chem. Abstr.*, **79**, 25244b (1973).

64   Tsukerman, S. V., Ch'ang, K-S., and Lavrushin, V. F., *Khim. Geterotsikl. Soedin., Akad. Nauk Latv. SSR (4)*, 537 (1965); *Chem. Abstr.*, **64**, 559g (1966).

65   Tsukerman, S. V., Ch'ang, K-S., and Lavrushin, V. F., *Zh. Prikl. Spectroskopii, Akad. Nauk Belorussk. SSR*, **4** (6), 554 (1966); *Chem. Abstr.*, **65**, 14660c (1966).

66   Lavrushin, V. F., Verkhovod, V. M., Pristupa, V. K., and Buryakovskaya, E. G., *Izv. Vyssh. Ucheb. Zaved., Khim. Khim. Tekhnol.*, **18**, 1504 (1975); *Chem. Abstr.*, **84**, 23912x (1976).

67   Tsukerman, S. V., Nikitchenko, V. M., and Lavrushin, V. F., *Zh. Obshch. Khim.*, **33**, 1255 (1963).

68   Tsukerman, S. V., Orlov, V. D., and Lavrushin, V. F., *Khim. Str., Svoistva Reaktivnost Org. Soedin.*, 63 (1969); *Chem. Abstr.*, **72**, 99698e (1970).

69   Roshka, V. K., *Tezisy Dokl. Soobshch.-Konf. Molodykh Uch. Mold.*, 9th (1974). (Pub. 1975), 106. Edited by A. M. Lazarev, "Shtiintsa": Kishinev USSR; *Chem. Abstr.*, **84**, 179102u (1976).

70   Rakosi-David, E., Bognár, R., and Tokes, A., *Acta Phys. Chim., Debrecina*, **14**, 145 (1968); *Chem. Abstr.*, **71**, 61713x (1969).

71   Surov, Y. N., Shkumat, H. P., Nikitchenko, V. M., Tsukerman, S. V., and Lavrushin, V. F., *Zh. Obshch. Khim.*, **48**, 2291 (1978).

72   Tolmachev, V. N., Boberov, O. F., and Lavrushin, V. F., *Dopov. Akad. Nauk Ukr. RSR* (11), 1454 (1966); *Chem. Abstr.*, **66**, 60470a (1967).

73   Rappoport, Z., and Sheradsky, T., *J. Chem. Soc., B*, 898 (1967).

74   Roberts, J. D., and Green, C., *J. Am. Chem. Soc.*, **68**, 214 (1946).

75   Johnson, G. D., *J. Am. Chem. Soc.*, **75**, 2720 (1953).

76   Yaroslavsky, S., *J. Org. Chem.*, **25**, 480 (1960).

77   Lavrushin, V. F., Grin, L. M., Pivnenko, N. S., and Kutsenko, L. M., *Ukr. Khim. Zh.*, **38**, 798 (1972); *Chem. Abstr.*, **78**, 3416m (1973).

78   Lecomte, J., and Guy, J., *Compt. Rend.*, **227**, 54 (1948).

79   Scrocco, M., and Liberti, A., *Ricerca Sci.*, **24**, 1687 (1954); *Chem. Abstr.*, **49**, 7979f (1955).

80   Kovalev, I. P., and Titov, E. V., *Zh. Obshch. Khim.*, **33**, 1670 (1963).

81 Sabata, B. K., and Rout, M. K., *J. Indian Chem. Soc.*, **41**, 74 (1964).

82 Mahanthy, P., Panda, S. P., Sabata, B. K., and Rout, M. K., *Indian J. Chem.*, **3**, 121 (1965).

83 Trakroo, P. P., and Mukhedkar, A. J., *J. Univ. Poona Sci. Technol.*, **28**, 101 (1964); *Chem. Abstr.*, **64**, 160g (1966).

84 Magdeeva, R. K., and Belotsvetov, A. V., *Izv. Vyssh. Ucheb. Zaved., Khim. Khim. Tekhnol.*, **13**, 794 (1970); *Chem. Abstr.*, **73**, 103778w (1970).

85 Dhar, D. N., and Gupta, V. P., *Indian J. Chem.*, **9**, 818 (1971).

86 Zalukaev, L. P., Vorobéva, R. P., and Oleinikova, T. A., *Zh. Prikl. Spektrosk.*, **21**, 558 (1974); *Chem. Abstr.*, **82**, 36867v (1975).

87 Dhar, D. N., and Misra, S. S., *J. Indian Chem. Soc.*, **49**, 629 (1972).

88 Misra, S. S., *Monatsh. Chem.*, **104**, 11 (1973).

89 Tsukerman, S. V., Orlov, V. D., Rozum, Y. S., and Lavrushin, V. F., *Khim. Geterotsikl. Soedin.*, **4**, 623 (1969); *Chem. Abstr.*, **72**, 26674j (1970).

90 Verkhovod, V. M., Khashchina, M. V., Berestetskaya, V. D., and Lavrushin, V. F., *J. Structural Chem.*, **18**, 26 (1977).

91 Verkohovod, V. M., Roberman, A. I., Ostrovskaya, B. I., and Lavrushin, V. F., *Vestn. Kharkov Un-ta* (175), 95 (1978); *Chem. Abstr.*, **91**, 55781q (1979).

92 Weber, F. G., Holzenger, A., and Westphal, G., *Z. Chem.*, **16**, 19 (1976).

93 Tsukerman, S. V., Surov, Y. N., and Lavrushin, V. F., *Zh. Obshch. Khim.*, **38**, 2411 (1968).

94 Yamaguchi, M., Hayashi, Y., and Matsukawa, S. *Bunseki Kagaku*, **10**, 1106 (1961); *Chem. Abstr.*, **56**, 11086i (1962).

95 Weber, F. G., and Brosche, K. C., *Z. Chem.*, **15**, 187 (1975).

96 Hergert, H. L., and Kurth, E. F., *J. Am. Chem. Soc.*, **75**, 1622 (1953).

97 Yanovskaya, L. A., Umirzakov, B., Yakovlev, I. P., and Kucherov, V. F., *Izv. Akad. Nauk SSSR, Ser. Khim.* (12), 2666 (1972); *Chem. Abstr.*, **78**, 110045m (1973).

98 Silver, N. L., and Boykin, D. W., Jr., *J. Org. Chem.*, **35**, 759 (1970).

99 Chawla, H. M., Chibber, S. S., and Seshadri, T. R., *Curr. Sci.*, **44**, 344 (1975); *Chem. Abstr.*, **83**, 42509s (1975).

100 Arventiev, B., Strat, G., Cascaval, A., and Strat, M., *An. Stiint, Univ. "Al. I. Cuza" Iasi, Sect. 1b*, **23**, 71 (1977); *Chem. Abstract.*, **91**, 4725s (1979).

101 Trakroo, P. P., and Mukhedkar, A. J., *J. Indian Chem. Soc.*, **41**, 595 (1964).

102 Beaupere, D., Uzan, R., and Doucet, J. P., *C. R. Acad. Sci., Ser. C*, **278**, 187 (1974); *Chem. Abstr.*, **80**, 132283f (1974).

103 Tsukerman, S. V., Surov, Y. N., and Lavrushin, V. F., *Zh. Obshch. Khim.*, **37**, 364 (1967).

104 Trusevich, N. D., Tolmachev, V. N., and Lavrushin, V. F., *Vestn. Kharkov Un-t, Khim.*, **127**, 98 (1975); *Chem. Abstr.*, **84**, 134982q (1976).

105 Perjessy, A., *Chem. Zvesti*, **23**, 441 (1969); *Chem. Abstr.*, **72**, 37296p (1970).

106 Kanate, B., Trusevich, N. D., Surov, Y. N., and Nikitchenko, V. M., *Vestn. Kharkov Un-t, Khim.*, **127**, 113 (1975); *Chem. Abstr.*, **84**, 163767d (1976).

107   Grouiller, A., Thomassery, P., and Pacheco, H., *Bull. Soc. Chim. (France)*, **12** (Part 2), 3452 (1973); *Chem. Abstr.*, **81**, 12960y (1974).

108   Hayes, W. P., and Timmons, C. J., *Spectrochim. Acta*, **24A**, 323 (1968).

109   Perjessy, A., *Chem. Zvesti*, **23**, 905 (1969); *Chem. Abstr.*, **74**, 47553a (1971).

110   Bellamy, L. J., *Infrared Spectra of Complex Molecules*, Methuen, London, 1958.

111   Bellamy, L. J., *Spectrochim. Acta*, **16**, 30 (1960).

112   Tsukerman, S. V., Surov, J. N., and Lavrushin, V. F., *Zh. Obshch. Khim.*, **38**, 524 (1968).

113   Favorskaya, I. A., and Plekhotkina, M. M., *Zh. Organ. Khim.*, **5**, 840 (1969); *Chem. Abstr.*, **71**, 38136c (1969).

114   Arbuzov, B. A., Yuldasheva, L. K., Anonimova, I. V., Shagidullin, R. R., Chernova, A. V., and Fazliev, D. F., *Izv. Akad. Nauk, SSSR, Ser. Khim.* (6), 1258 (1969); *Chem. Abstr.*, **71**, 80535g (1969).

115   Dzurilla, M., and Kristian, P., *Coll. Czech. Chem. Commun.*, **35**, 417 (1969).

116   Dzurilla, M., Kristian, P., and Gyoryova, K., *Chem. Zvesti*, **24**, 207 (1970); *Chem. Abstr.*, **74**, 53196c (1971).

117   Dinya, Z., and Litkei, G., *Acta Chim. Acad. Sci. Hung.*, **75**, 161 (1973).

118   Tsukerman, S. V., Izvekov, V. P., Rozum, Y. S., and Lavrushin, V. F., *Khim. Geterotsikl. Soedin.* (6), 1011 (1968); *Chem. Abstr.*, **70**, 72307y (1969).

119   Kohlrausch, K. W. F., and Pongratz, A., *Monatsh. Chem.*, **64**, 374 (1934).

120   Murti, G. V. L. N., and Seshadri, T. R., *Proc. Indian Acad. Sci.*, **8A**, 519 (1938).

121   Harrand, M., and Guy, J., *Compt. Rend.*, **226**, 480 (1948).

122   Mabry, T. J., Markham, K. R., and Thomas, M. B., *The Systematic Identification of Flavonoids*, Springer-Verlag, New York, Heidelberg, Berlin, 1970, pp. 335–339.

123   Tsukerman, S. V., Orlov, V. D., Yatsenko, A. I., and Lavrushin, V. F., *Teor. Eksp. Khim.*, **6**, 67 (1970); *Chem. Abstr.*, **73**, 34407k (1970).

124   Grin, L. M., Pivnenko, N. S., Kutsenko, L. M., and Lavrushin, V. F., *Zh. Organ. Khim.*, **6**, 1904 (1970); *Chem. Abstr.*, **73**, 135762c (1970).

125   Membery, F., and Doucet, J. P., *J. Chim. Phys.*, **73**, 1024 (1977).

126   Radeglia, R., Orlov, V. D., and Reinhardt, M., *Z. Chem.*, **17**, 377 (1977).

127   Nesmeyanov, A. N., Shul'pin, G. B., Fedorov, L. A., Petrovsky, P. V., and Rybinskaya, M. I., *J. Organometal. Chem.*, **69**, 429 (1974).

128   Holak, F. A., Kawalek, B., and Mirek, J., *Bull. Acad. Pol. Sci., Ser., Sci. Chim.*, **26**, 449 (1978); *Chem. Abstr.*, **89**, 214551s (1978).

129   Reichel, L., and Neubauer, A., *Z. Chem.*, **8**, 180 (1968).

130   Balaban, A. T., Zugrăvescu, I., Avramovici, S., and Silhan, W., *Monatsh. Chem.*, **101**, 704 (1970).

131   Beynon, J. H., Lester, G. R., and Williams, A. E., *J. Phys. Chem.*, **63**, 1861 (1959).

132   Bowie, J. H., Grigg, R., Williams, D. H., Lawesson, S. O., and Schroll, G., *Chem. Commun.* (17), 403 (1965).

133 Van De Sande, C., Serum, J. W., and Vandewalle, M., *Org. Mass Spectrom.*, **6**, 1333 (1972).

134 Jackson, B., Locksley, H. D., Scheinmann, F., and Wolstenholme, W. A., *Tetrahedron Letters*, 3049 (1967).

135 Itagaki, Y., Kurokawa, T., Sasaki, S., Chang, C-T., and Chen, F-C., *Bull. Chem. Soc. Japan*, **39**, 538 (1966).

136 Dinya, Z., Litkei, G., Tamas, J., and Czira, G., *Acta Chim. Acad. Sci. Hung.*, **84**, 181 (1975); *Chem. Abstr.*, **82**, 155001t (1975).

137 Chizhov, O. S., Shamshurina, S. A., Zolotarev, B. M., Yanovskaya, L. A., and Umirzakov, B., *Izv. Akad. Nauk SSSR, Ser. Khim.* (6), 1389 (1973); *Chem. Abstr.*, **79**, 104471b (1973).

Chapter Twenty-Four

Chromatographic Separation of Chalcones from Other Flavonoids

COLUMN CHROMATOGRAPHY

Column chromatography over silica gel, involving the use of berberine as a fluorescent adsorbent, has been employed[1] successfully for the separation of chalcone from 4-methoxychalcone.

2'-Hydroxychalcones are separable from the corresponding flavanones, for example, butin from butein, by column chromatography, using polyamide[2] as adsorbent. 2'-Hydroxychalcones are retained by the column, while flavanones are eluted with aqueous methanol. The chalcones, adsorbed on the column, are then eluted with the pure solvent. The separation of 2'-hydroxychalcones from flavonols[3] and flavanonols[3] have likewise been carried out by polyamide column chromatography.

196

THIN-LAYER CHROMATOGRAPHY

Thin-layer chromatography (TLC) on silica gel or polyester film has been used for effecting the separation of chalcones from other flavonoids and alkaloids.[4] Hesperidin methylchalcone in pharmaceuticals has been isolated on polyamide thin layers and subsequently estimated by fluorometric method (treatment of the chalcone spot with methanolic aluminum chloride, and the fluorescent spot removed, extracted with a solvent, and subjected to spectrophotometric analysis).[5]

A preparative TLC method has been developed, for separation on silica gel, for several hydroxychalcones and their corresponding flavanones, using ligroin–ethylacetate as the developer solvent.[6]

The *cis–trans* isomers of 2-hydroxy-α-methoxychalcone have been separated on silica gel plates, using eluotropic solvent, benzene–methanol (95:5).[7]

R_f values are reported for a large number of substituted aromatic chalcones,[8-10] ferrocene,[11] furyl,[10] thienyl,[9,10] and pyrryl[9] analogues of chalcones. The R_f values have been correlated with the position and nature of the substituents.[8]

PAPER CHROMATOGRAPHY

Chalcones have been separated from other flavonoids by paper chromatography,[12-14] using butanol–acetic acid–water (4:1:2.2) as the developer solvent. The spot positions of chalcones on the chromatogram are revealed either by characteristic fluorescence in the ultraviolet light or by color reactions.

Paper chromatography of some chalcones[15,16] and their glycosides[15] are reported. These give yellow or orange spots on paper developed with aqueous phenol.[15]

Paper chromatography, involving the use of boric acid–sodium acetate impregnated paper, has been utilized for differentiating *o*-dihydroxyflavanones from similar flavanones, but which are lacking in catechol-type hydroxylic grouping.[17]

The relation between chromatographic parameters of halogenated chalcones by adsorption and partition chromatographic techniques has been studied.[18]

ION-EXCHANGE CHROMATOGRAPHY

Chalcone is reported to be quantitatively adsorbed on the anion exchange resin (Amerlite IRA-400, in HSO_3^- form), provided the passage through the column is long and can be eluted with sodium chloride solution.[19]

REFERENCES

1   Brockmann, H., and Volpers, F., *Chem. Ber.*, **82**, 95 (1949).

2   Neu, R., *Nature*, **182**, 660 (1958).

3   Neu, R., *Arch. Pharm.*, **293**, 169 (1960); *Chem. Abstr.*, **550**, 24734c (1961).

4   Paris, R. R., Rousselet, R., Paris, M., and Fries, J., *Ann. Pharm. Franc.*, **23**, 473 (1965); *Chem. Abstr.*, **64**, 6404d (1966).

5   Silvestri, S., *Pharm. Acta Helv.*, **45**, 390 (1970); *Chem. Abstr.*, **73**, 102100g (1970).

6   Dhar, D. N., *J. Indian Chem. Soc.*, **49**, 309 (1972).

7   Pastuska, G., Petrowitz, H. J., and Krueger, R., *Fresenius Z. Anal. Chem.*, **236**, 333 (1968).

8   Dhar, D. N., Tewari, R. P., Ahuja, A. P., and Tripathi, R. D., *J. Proc. Inst. Chem. (India)*, **43**, 198 (1971).

9   Dhar, D. N., *J. Chromatog.*, **67**, 186 (1972).

10   Dhar, D. N., and Misra, S. S., *J. Chromatog.*, **69**, 416 (1972).

11   Bozak, R. E., and Fukuda, J. H., *J. Chromatog.*, **26**, 501 (1967).

12   Mikhailov, M. K., *Doklady Akad. Nauk SSSR*, **108**, 511 (1956); *Chem. Abstr.*, **50**, 17342h (1956).

13   Nikonov, G. K., *Med. Prom. SSSR*, **12**, 16 (1959); *Chem. Abstr.*, **53**, 14094b (1959).

14   Curtze, A., and Holland, W., *Dsch. Apoth.-Ztg.*, **107**, 147 (1967); *Chem. Abstr.*, **66**, 108283w (1967).

15   Puri, B., and Seshadri, T. R., *J. Sci. Ind. Research (India)*, **12B**, 462 (1953).

16   Fujise, S., and Tatsuta, H., *J. Chem. Soc. Japan*, **73**, 35 (1952).

17   Jurd, L., *J. Chromatog.*, **4**, 369 (1960).

18   Covello, M., Schettino, O., Ferrara, L., and Forgione, P., *Rend. Accad. Sci. Fis. Mat., Naples*, **41**, 290 (1975); *Chem. Abstr.*, **84**, 112062n (1976).

19   Sjöström, E., *Acta Polytech.*, **144**, 7 (1954); *Chem. Abstr.*, **49**, 1483i (1955).

PART FOUR

Applications

Chapter Twenty-Five

Naturally Occurring Chalcones and Their Derivatives

During the past decade a number of reports have appeared in the literature, describing the isolation of chalcones from various parts of plants: roots,[6,19,20,24,82,83] heartwood,[44,73,79,80] buds,[4] leaves,[23,27,29,34,36,51,52,54,58,71] blossoms,[45–47] inflorescences,[33] flowers,[2,8,17,18,23,29,42] and seeds.[5,53,57,71,74] These compounds exist in the free state as chalcones or in the combined form as glycosides. The substituent(s), hydroxy, methoxy, methyl, and isopentenyl, may be present either in ring A and/or ring B of the chalcone molecule. Moreover, there are reports[58,66,70,76,82–85] about the isolation of dihydrochalcones from the higher plants.

Table 1 lists the naturally occurring chalcones and their derivatives.

Table 1 Naturally Occurring Chalcones and Their Derivatives

Source	Chalcones and Their Derivatives	Ref.
Acacia auriculiformis	2′,3′,4′,4-Tetrahydroxychalcone	1
Acacia cyanophylla (flowers)	Chalcononaringenin-4-glucoside; isosalipurposide	2
Adenanthera pavonina "red sandalwood"	Chalcone; 2′,3,4,4′-tetrahydroxychalcone (butein)	3
Alnus virdis (buds)	4′,5′-Dihydroxy-6′-methoxychalcone	4
Alpinia speciosa (seeds)	2′,4′-Dihydroxy-6′-methoxychalcone (cardamonin)	5
Angelica Keiskei (roots)	Prenylchalcones: xanthoangelol; hydroxyderricin	6
Aniba rosaeodora Ducke	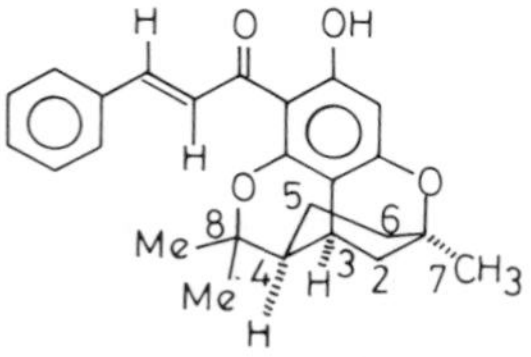	7
Antirrhinum majus (yellow flowers)	Chalcononargenin-4′-glucoside; 2′,3,4,4′,6′-pentahydroxychalcone-4′-glucoside	8
Berchemia zeyheri "red ivory" (redwood)	α,2′,4,4′,6′-Pentahydroxychalcone	9
Bidens tripartite	2,3′,4,4′-Tetrahydroxychalcone; butein-7-*O*-β-D-glucopyranoside	10
Chromalaena odorata	2′,4-Dihydroxy-4′,5′,6′-trimethoxychalcone; 2′-hydroxy-4,4′,5,5′,6′-pentamethoxychalcone	11
"Citrus molasses"	A methylchalcone	12
Cordoa piaca (lonchocarpus)	Cordoin; isocordoin; ψ-isocordoin, dihydrocordoin; derricin, 4-hydroxyderricin; lonchocarpin; 4-hydroxyloncho-carpin and 4-hydroxyisocordoin	13–16
Coreopsis tinctoria (ray flowers)	4′-Glucosidoxy-2′,3′,3,4-tetrahydroxychalcone (marein)	17
"Cotton" (flowers)	2′,3′,4′,6′,8′,3,4-Heptahydroxychalcone glucoside	18

Table 1 **(Continued)**

Source	Chalcones and Their Derivatives	Ref.
Cryptocarya *bourdilloni* (Lauraceae) (roots)		19,20
Daemonorops dracu "dragon's blood resin" (Fruit)	2,4-Dihydroxy-5-methyl-6-methoxy-chalcone; 2,4-dihydroxy-6-methoxy-chalcone	21
Dahlia species	4'-Arabinosylgalactoside of butein and other glycosides	22
Dahlia tenuicaulis (leaves)	2'-hydroxy-4,4',6'-trimethoxychalcone	23
Dahlia tenuicaulis (flower heads)	4,2',4'-Trihydroxychalcone; 3,2',4'-tri-hydroxy-4-methoxychalcone	23
Datisca cannabina (root cortex)	Unidentified chalcone	24
Derris sericea (root bark)	$R = (Me_2C=CH-CH_2-)$	25
Didymocarpus *pendicellata* (Gesneriaceae) (roots)	2',6'-Dihydroxy-4',5'-dimethoxychal-cone (pashanone)	26
Eupatorium odoratum (leaves)	2'-Hydroxy-4,4',5',6'-tetramethoxychal-cone	27
Flemingia Chappar Ham. (whole plant)	2',4'-Dihydroxychalcone; 2',4',4-trihydroxychalcone; 2',4'-dihydroxy-5'-methoxychalcone;	28–32
Flemingia Chappar Ham. (flower/leaves)	4',6'-Dihydroxy-3'-methoxychalcone	29

Table 1 (Continued)

Source	Chalcones and Their Derivatives	Ref.
Flemingia congesta (inflorescences)	(Chromenochalcone)	33
Flemingia stricta (Leguminoseae) (leaves)	Flemistrictin—A	34
Flemingia strobilifera (roots)	3′,6′-Dihydroxy-2′,4′,5′,4-tetramethoxy-chalcone	35
Flemingia wallichii (leaves)	Homoflemingin; flemiwallichin	36
Glycyrrhiza echinata (tissue culture)	4,4′-Dihydroxy-2-methoxychalcone (echinatin)	37
Glycyrrhiza glabra (roots)	2,4,4′-Trihydroxychalcone; 4-hydroxychalcone, unidentified chalcones; glycosidal chalcone	38,39
Glycyrrhiza glabra (root bark)	(I) R = (CMe$_2$=CH-CH$_2$); R′ = H (II) R = H; R′ = OH	41

Table 1 **(Continued)**

Source	Chalcones and Their Derivatives	Ref.
Gnaphalium affine (flowers)		42
Gnaphalium multiceps		43
Goniorrhachis marginata (heartwood)		44
Helichrysum bracteatum (blossoms)	2',3,4,4',6'-Pentahydroxychalcone-2'-glucoside; isosalipurposide; 2,3,4,4',5,6'-hexahydroxychalcone-2'-glucoside	45–47
"Hop extract"	 (Xanthohumol)	48
Larrea nitida (aerial parts)	2',4'-Dihydroxy-3'-methoxychalcone (larrein); 2',4'-dihydroxychalcone	49
Lasthenia (Compositae)	Butein and okanin	50
Lindera umbellata Thumb (leaves)	2',6'-Dihydroxy-4'-methoxychalcone	51–52
Lonchocarpus sericeus (Belgian Congo) (seeds and roots)		53

205

Table 1   **(Continued)**

Source	Chalcones and Their Derivatives	Ref.

Lyonia formosa (leaves)	(lyonogenin); lyonogenin-2'-glucopyranoside (lyonotin)	54
Machaerium nucronulatum (wood)	Butein isoliquiritigenin	55
Merrillia caloxylon (fruit)	2'-Hydroxy-3,4,4',6'-tetramethoxychalcone; 2',3-dihydroxy-4,4',6'-trimethoxychalcone; 2-hydroxy-3,4,4',5',6'-pentamethoxychalcone	56
Milletia ovalifolia (seeds)	2'-Hydroxy-3'-C-prenyl-4',6'-dimethoxychalcone (ovalichalcone)	57

(Ovalitenin A)

(Ovalitenin B)

Myrica gale (leaves)	2',4'-Dihydroxy-6'-methoxy-3',5'-dimethylchalcone	58
Myrica gale (fruits)	2'-Hydroxy-4',6'-dimethoxy-3'-methyldihydrochalcone	58
Myrica gale (fruits)	2',6'-Dihydroxy-4'-methoxy-3',5'-dimethyldihydrochalcone;	59

Table 1 (Continued)

Source	Chalcones and Their Derivatives	Ref.
Oenothera hookeri	Isosalipurposide	60
Onagraceae (petals)	Onagraceae: 13 species contain the chalcone, while 27 species are lacking in it	61
Onychium auratum (fern)	2',6'-Dihydroxy-4',5'-dimethoxychalcone; 2',6'-dihydroxy-4'-methoxychalcone	63
Petnuia hybrida (pollen)	4,2',4',6'-Tetrahydroxychalcone	64
Piper methysticum Forst (Roots)	2',4'-Dihydroxy-4,6-dimethoxychalcone	65
Pityrogramma austroamericana (gold fern) (Yellow powdery coating on the underside of fronds)	2',6'-Dihydroxy-4,4'-dimethoxychalcone; 2',6'-dihydroxy-4,4'-dimethoxydihydrochalcone	66
Pityrogramma calomelanos; *P. tarterea* (exudate) (Jamaican ferns)	2',6'-Dihydroxy-4'-methoxydihydrochalcone; 2',6'-dihydroxy-4,4'-dimethoxydihydrochalcone	67
Pityrogramma chrysophylla heyderi (fronds)	2',6'-Dihydroxy-4'-methoxychalcone; 2',6'-dihydroxy-4,4'-dimethoxychalcone	66 68
Pityrogramma lehmanii (silver fern) (white coating on the underside of fronds)	2',6'-Dihydroxy-4,4'-dimethoxydihydrochalcone	66
Pityrogramma triangularis (ferns)(exudate)	2',6'-Dihydroxy-3'-methyl-4'-methoxychalcone	69
Podocarous nubigena	α,2',4,4',6'-Pentahydroxy-dihydrochalcone (nubigenol)	70
Polygonum senegalese (Seeds and leaves)	2',4'-Dihydroxy-3',6'-dimethoxychalcone	71
"Populus bud oils" (several species or hybrids of populus)	2',4',6'-Trihydroxychalcone; 2',6'-dihydroxy-4'-methoxychalcone	72
Prunus cerasus L. (heartwood)	2'-Hydroxy-2,4,4',6'-tetramethoxychalcone (cerasidin); 2',4'-dihydroxy-2,4,6'-trimethoxychalcone (cerasin)	73

Table 1 (Continued)

Source	Chalcones and Their Derivatives	Ref.

Psorelea corylifolia (seeds)	Isobavachalcone; 4'-*O*-methylchalcone; 5'-formyl-2',4-dihydroxy-4'-methoxy-chalcone	74 75
Rhododendron canescens; R. nudiflorum; R. roseum	2',4,6'-Trihydroxy-4'-methoxydihydro-chalcone; asebotin	76
Salix acutifolia (bark)	2',4,4',6'-Tetrahydroxychalcone-6'-(6-*O*-*p*-coumaroyl)-D-glucopyranoside; chalconaringenin-6'-D-glucoside	77
Tephrosia obovata Merr. (fish poison plant)	Pyranochalcone	78

Trechylobium verrucosum (Gaertn.) oliv. (heartwood)	α,2',3,4,4'-Pentahydroxychalcone	79,80
Tulipa CV. "Apeldoorn" (during development of anthers)	2',3,4,4',6'-Pentahydroxychalcone; 2',4,4',6'-Tetrahydroxy-3-methoxychal-cone; 2,4,4',6-Tetrahydroxychalcone	81
Uvaria acuminata (roots)		82

Uvaria chamae (roots)	Benzyldihydrochalcones; chamuvarin; chamuvaritin	83
Viburnum davidii	2',4,4'-Trihydroxydihydrochalcone	84
Viburnum davidii (leaves)	Davidioside; davidigenin; 4'-methoxy derivative	85
Viseum album L.	2'-Hydroxy-4',6'-dimethoxychalcone	86

REFERENCES

1   Drewes, S. E., and Roux, D. G., *Biohem. J.*, **98**, 493 (1966).

2   Imperato, F., *Phytochemisry*, **17**, 822 (1978).

3   Gennaro, A., Merlini, L., and Nasini, G., *Phytochemistry*, **11**, 1515 (1972).

4   Woolenweber, E., Jay, M., and Favre-Bonvin, J., *Phytochemistry*, **13**, 2618 (1974).

5   Krishena, B. M., and Chaganty, R. B., *Phytochemistry*, **12**, 238 (1973).

6   Kozawa M., Morita, N., Baba, K., and Hata, K., *Yakugaku Zasshi*, **98**, 210 (1978); *Chem. Abstr.*, **88**, 191090d (1978).

7   Combes, G., Vassort, P., and Winternitz, F., *Tetrahedron*, **26**, 5981 (1970).

8   Gilbert, R. I., *Phytochemistry*, **12**, 809 (1973).

9   Volsteedt, F. du R., Rall, G. J. H., and Roux, D. G., *Tetrahedron Letters*, 1001 (1973).

10 Serbin, A. G., Borisov, M. I., and Chernobai, V. T., *Khim. Prir. Soedin.*, **8**, 440 (1972); *Chem. Abstr.*, **78**, 1987z (1973).

11 Barua, R. N., Sharma, R. P., Thyagarajan, G., and Hertz, W., *Phytochemistry*, **17**, 1807 (1978).

12 Sokoloff, B. T., U.S. Patent 2,734,896 (1956); *Chem. Abstr.*, **50**, 7408a (1956).

13 Goncalves de Lima, O., Marini-Bettolo, G. B., Francisco de Mello, J., Delle Monache, F., Coelho, J. S. de B., Lyra, F. D. de A., and Fernandes de Albuquerque, M. M., *Atti. Accad. Naz. Lincei, Cl. Sci. Fis., Mat. Nat., Rend.*, **53**, 433 (1973); *Chem. Abstr.*, **80**, 70483k (1974); **82**, 1915g (1975).

14 Francisco de Mello, J., Delle Monache, F., Goncalves de Lima, O., Marini-Bettolo, G. B., Delle Monache, G., Lyra, F. D. de A., Fernandes de Albuquerque, M. M., and Leoncio d'Albuquerque, I., *Rev. Inst. Antibiot. Univ. Fed. Pernambuco, Recife*, **13**, 37 (1973); *Chem. Abstr.*, **82**, 167508b (1975).

15 Delle Monache, G., Francisco de Mello, J., Delle Monache, F., Marini-Bettolo, G. B., Concalves de Lima, O., and Coelho, J. S. de B., *Gazz. Chim. Ital.*, **104**, 861 (1974).

16 Goncalves de Lima, O., Francisco de Mello, J., Coelho, J. S. de B., Lyra, F. D. de A., Fernandes de Albuquerque, M. M., Marini-Bettolo, G. B., Delle Monache, G., and Delle Monache, F., *Farmaco, Ed. Sci.*, **30**, 326 (1975); *Chem. Abstr.*, **83**, 91495t (1975).

17 Shimokoriyama, M., *J. Am. Chem. Soc.*, **79**, 214 (1957).

18 Pakudina, Z. P., Rakhimov, A. A., and Sadykov, A. S., *Khim. Prir. Soedin* (2), 126 (1969); *Chem. Abstr.*, **71**, 88425s (1969).

19 Govindachari, T. R., and Parthasarthy, P. C. *Tetrahedron Letters*, 3419 (1972).

20 Govindachari, T. R., Parthasarthy, P. C., Desai, H. K., and Shanbhag, M. N., *Tetrahedron*, **29**, 3091 (1973).

21 Cardillo, G., Merlini, L., Nasini, G., and Salvadori, P., *J. Chem. Soc. C*, 3967 (1971).

22 Giannasi, D. G., *Mem. N.Y. Bot. Gard.*, **26**, 125 (1975).

23 Laur, J., and Wrang, P., *Phytochemistry*, **14**, 1621 (1975).

24 Zapesochnaya, G. G., Ban'kovskii, A. I., and Molodozhnikov, M. M., *Khim. Prir. Soedin.*, **5**, 179 (1969); *Chem. Abstr.*,'**71**, 98962d (1969).

25 Do-Nascimanto, M. C., and Mors, W. B., *Phytochemistry*, **11**, 3023 (1972).

26 Agarwal, S. C., Bhaskar, A., and Seshadri, T. R., *Indian J. Chem.*, **11**, 9 (1973).

27 Bose, P. K., Chakrabarti, P., Chakrabarti, S., Dutta, S. P., and Barua, A. K., *Phytochemistry*, **12**, 667 (1973).

28 Adityachaudhury, N., Kirtaniya, C. L., and Mukherjee, B., *J. Indian Chem. Soc.*, **46** 964 (1969).

29 Cardillo, B., Gennaro, A., Merlini, L., Nasini, G., and Servi, S., *Tetrahedron Letters*, 4367 (1970).

30 Adityachaudhury, N., Kirtaniya, C. L., and Mukherjee, B., *J. Indian Chem. Soc.*, **47**, 508 (1970).

31 Adityachaudhury, N., Kirtaniya, C. L., and Mukherjee, B., *Tetrahedron*, **27**, 211 (1971).

32 Adityachaudhury, N., and Kirtaniya, C. L., *J. Indian Chem. Soc.*, **47**, 1023 (1970).

33 Cardillo, B., Gennaro, A., Merlini, L., Nasini, G., and Servi, S., *Phytochemistry*, **12**, 2027 (1973).

34 Rao, J. M., Subrahmanyam, K., and Rao, K. V. J., *Curr. Sci. (India)*, **44**, 158 (1975).

35 Bhatt, S., *Indian J. Chem.*, **13**, 1105 (1975).

36 Rao, J. M., Babu, S. S., Subrahmanyam, K., and Rao, K. V. J., *Curr. Sci.*, **47**, 584 (1978).

37 Furuya, T., Matsumoto, K., and Hikichi, M., *Tetrahedron Letters*, 2567 (1971).

38 Litvinenko, V. I., Maksyutina, N.P., and Kolesnikov, I. G., *Zh. Obshch. Khim.*, **33**, 296 (1963).

39 Horton-Dorge, M., *J. Pharm. Belg.*, **29**, 560 (1974); *Chem. Abstr.*, **83**, 111099x (1975).

40 Pointet-Guillot, M., Brit. Patent, 901,086 (1962); *Chem. Abstr.*, **57**, 11318a (1962).

41 Saitoh, T., and Shibata, S., *Tetrahedron Letters*, 4461 (1975).

42 Aritomi, M., and Kawasaki, T., *Chem. Pharm. Bull.*, **22**, 1800 (1974); *Chem. Abstr.*, **81**, 166401d (1974).

43 Maruyama, M., Hayasaka, K., Sasaki, S., Hosokawa, S., and Uchiyama, H., *Phytochemistry*, **13**, 286 (1974).

44 Gottlieb, O. R., and Regode Sousa, J., *Phytochemistry*, **11**, 2841 (1972).

45 Kaufmann, H. P., El Baya, Abd El W., *Fette, Seifen, Anstrichm.*, **71**, 25 (1969).

46 Rimpler, H., Langhammer, L., and Frenzel, H. J., *Planta Med.*, **110,** 325 (1963).

47 Rimpler, H., and Haensel, R., *Arch. Pharm.*, **298**, 838 (1965); *Chem. Abstr.*, **64**, 13015h (1966).

48 Verzele, M., Stockx, J., Fontijn, F., and Anteunis, M., *Bull. Soc. Chim. Belges*, **66**, 452 (1957); *Chem. Abstr.*, **52**, 7243d (1958).

49 Federiva, R., Kavka, J., and D'Arcangelo, A. T., *An. Asoc. Quim. Argent.*, **63**, 85 (1975); *Chem. Abstr.*, **83**, 203739j (1975).

50   Bohm, B. A., Saleh, N. A. M., and Ornduff, R., *Amer. J. Bot.*, **61**, 551 (1974).

51   Hayashi, N., Takeshita, K., Nishio, N., and Hayashi, S., *Chem. Ind. (London)*, **49**, 1779 (1969).

52   Nanao, H., Kenji, T., Noamichi, N., and Shuichi, H., *Flavour Ind.*, (6), 405 (1970); Chem. Abstr., **73**, 59214h (1970).

53   Baudrenghien, J., Jadot, J., and Huls, R., *Bull. Classe Sci., Acad. Roy. Belg.*, **39**, 105 (1953); *Chem. Abstr.*, **48**, 10744d (1954).

54   Shukla, Y. N., Tandon, J. S., and Dhar, M. M., *Indian J. Chem.*, **11**, 720 (1973).

55   Kurosawa, K., Ollis, W. D., Sutherland, I. O., Gottlieb, O. R., and De Oliveria, A. B., *Phytochemistry*, **17**, 1389 (1978).

56   Fraser, A. W., and Lewis, J. R., *Phytochemistry*, **13**, 1561 (1974).

57   Gupta, R. K., and Krishnamurti, M., *Phytochemistry*, **16**, 293, 1104 (1977).

58   Malterud, K. E., Anthonsen, T., and Lorentzen, G. B., *Phytochemistry*, **16**, 1805 (1977).

59   Wyar, T., Malterud, K. E., and Anthonsen, T., *Phytochemistry*, **17**, 2011 (1978).

60   Dement, W. A., and Raven, P. H., *Phytochemistry*, **12**, 807 (1973).

61   Dement, W. A., and Raven, P. H., *Nature*, **252** (5485), 705 (1974).

62   Ramakrishnan, G., Banerji, A., and Chadha, M. S., *Phytochemistry*, **13**, 2317 (1974).

63   Banerji, A., Ramakrishnan, G., and Chadha, M. S., India, A. E. C., Bhabha At. Res. Cent. (Rep.), BARC-764, 14 (1974); *Chem. Abstr.*, **82**, 108863r (1975).

64   De Vlaming, P., and Kho, K. F. F., *Phytochemistry*, **15**, 348 (1976).

65   Dutta, C. P., Ray, L. P. K., Chatterjee, A., and Roy, D. N., *Indian J. Chem.*, **11**, 509 (1973).

66. Wollenweber, E., *Pflanzenphysiologie*, **78**, 344 (1976); *Chem. Abstr.*, **85**, 17060x (1976).

67   Star, A. E., and Mabry, T. J., *Phytochemistry*, **10**, 2817 (1971).

68   Nilsson, M., *Acta Chem. Scand.*, **15**, 211 (1961).

69   Star, A. E., Mabry, T. J., and Smith, D. M., *Phytochemistry*, **17**, 586 (1978).

70   Bhakuni, D. S., Bittner, M., Silva, M., and Sammes, P. G., *Phytochemistry*, **12**, 2777 (1973).

71   Maradufu, A., and Ouma, J. H., *Phytochemistry*, **17**, 823 (1978).

72   Wollenweber, E., and Weber, W., *Pflanzenphysiologie*, **69**, 125 (1973); *Chem. Abstr.*, **78**, 133406u (1973).

73   Nagarajan, G. R., and Parmar, V. S., *Phytochemistry*, **16**, 1317 (1977).

74   Bajwa, B. S., Khanna, P. L., and Seshadri, T. R., *Current Sci.*, **41**, 814 (1972).

75   Gupta, S. R., Seshadri, T. R., and Sood, G. R., *Indian J. Chem.*, **13**, 632 (1975).

76   Mabry, T. J., Sakakibara, M., and King, B., *Phytochemistry*, **14**, 1448 (1975).

77   Vinokurov, I. I., and Skrigan, A. I., *Vestsi Akad. Nauk. Belarus, SSR, Ser. Khim. Nauk* (5), 112 (1969); *Chem. Abstr.*, **72**, 55822m (1970).

78   Chen. Y-L., *Asian J. Pharm.*, **3**, 18 (1978); *Chem. Abstr.*, **90**, 203903c (1979).

79   Vander Merwe, J. P., Ferreira, D., Bradt, E. V., and Roux, D. G., *Chem. Commun.* (9), 521 (1972).

80   Ferreira, D., Vander Merwe, J. P., and Roux, D. G., *J. Chem. Soc., Perkin Trans.*, **12**, 1492 (1974).

81   Quast, L., and Wiermann, R., *Experientia*, **29**, 1165 (1973).

82   Cole, J. R., Torrance, S. J., Wiedhopf, R. M., Arora, S. K., and Bates, R. B., *J. Org. Chem.*, **41**, 1852 (1976).

83   Okorie, D. A., *Phytochemistry*, **16**, 1591 (1977).

84   Bohm, B. A., and Glennie, C. W., *Phytochemistry*, **8**, 905 (1969).

85   Jensen, S. R., Nielsen, B. T., and Nern, V., *Phytochemistry*, **16**, 2036 (1977).

86   Becker, H., Exner, J., and Schilling, C., *Z. Naturforsch.*, **33C**, 771 (1978).

Chapter Twenty-Six

Biologically Active Chalcones and Their Derivatives

INTRODUCTION

The presence of enone function in the chalcone molecule confers antibiotic activity[1-35] (bacteriostatic/bactericidal) upon it. This property is enhanced when substitution is made at the α (nitro- and bromo-) and β- (bromo- and hydroxylic-) positions.[27] Some substituted chalcones and their derivatives, including some of their heterocyclic analogues, have been reported to possess some interesting biological properties, which are detrimental to the growth of microbes,[23-26] tubercle bacilli,[30-35] malarial parasites,[37] acrus,[38] *Schistosoma*,[39] and intestinal worms.[25,40,41] Some of the compounds are claimed to be toxic to animals[30,46] and insects[42,43] and are also reported to exhibit inhibitory action on several enzymes,[53-61] fungi,[1,3,17,22,48-51] and herbaceous plants.[43,47] The compounds of the chalcone series also show a profound influence on the cardiovascular, cerebrovascular, and neuromuscular systems, including the vital organs of the experimental animals.[63-125] The data on the biological activities of these compounds are summarized in the following table.

Compound	Remarks	Test Organism	Ref.

BIOLOGICAL ACTIVITIES OF CHALCONES
Bacteriostatic or Bactericidal Activity

Compound	Remarks	Test Organism	Ref.
Chalcone	Significant bacteriostatic action.	E. coli S. aureus B. mycoides B. subtilis S. lutea	1
	The bacteriostatic effect could not be reversed by cystein, in contrast to the effect of this compound on other antibiotics.	S. aureus	2
	The antibacterial action is associated with α,β-unsaturated carbonyl group of the molecule.	S. aureus 209P E. coli T. mentagrophytes C. albicans W. anomala T. utilis A. usami P. chrysogenum Q 176 S. sake	3
Prenylated chalcone	4-Hydroxyderricin exhibits a marked inhibitory activity (in vitro).	Gram-positive microorganisms	4
Halohydroxychalcones	Chalcones have 4'-hydroxyl and halogenic substituents (in 2 and 4 positions) possess marked antibacterial activity. Also the chalcones with a fluoro substituent have better antibacterial activity compared to bromo- or chlorochalcones.	S. albus S. aureus	5
Nitrohydroxychalcone	3'-Nitro-4'-hydroxy-2-methoxychalcone has the highest antibacterial activity (in vivo). Other active chalcones are 3'-nitro-4'-hydroxy-2,3-	S. albus	6, 7

Compound	Remarks	Test Organism	Ref.

BIOLOGICAL ACTIVITIES OF CHALCONES

	dimethoxychalcone and 3'-nitro-4'-hydroxy-2,5-dimethoxychalcone.		
	The following chalcones exhibit antibacterial activity: $RCOCH{=}CH{-}C_6H_4OMe{-}m$; $R = $ 2,5-, 4,3-, 4,2-, and 2,4-(OH) $NO_2C_6H_3$.	*E. coli* *S. flexneri* *K. pneumoniae* *S. aureus* *S. albus* Strain-resistant to streptomycin and polymyxin	8
	3'-Nitro-4'-hydroxy-2-methoxychalcone has a strong antibacterial action.	*S. albus* (*in vitro*)	7
Aminochalcones	4- (and 4'-) Aminochalcones possess bacteriostatic activity.	*S. aureus* *S. hemolyticus*	9
Chlorohydroxychalcone	Antibacterial.	—	10, 11
Bromohydroxychalcone	Antibacterial.	*S. aureus*	12
	4',5-Dibromo-2-hydroxychalcone has the highest bacteriostatic activity (1:640,000).	—	13
	2,2'-Dihydroxy-3,5,5'-tribromochalcone possesses bacteriostatic action. Inhibitory concentration: 1:640,000.	*S. aureus*	14
Iodohydroxychalcone	4'-Hydroxy-3',4,5'-triiodochalcone and 4'-hydroxy-3',5'-diiodo-4-propoxy- (and 4-butoxy-) chalcones are considered as possible antibacterial agents.	—	15

Compound	Remarks	Test Organism	Ref.

BIOLOGICAL ACTIVITIES OF CHALCONES

Compound	Remarks	Test Organism	Ref.
Alkylthiochalcones	The following chalcones possess antibacterial activity (*in vitro*): 4'-thioalkylchalcone and 4-chloro-5'-methyl-2'-thioalkyl chalcone.	*B. subtilis* NRRL *E. coli* 0-55 *S. aureus* 209P	16
Sulfonic acid and carboxylic acid derivatives of chalcones	Bactericidal.	—	17
Chalcone sulfanilamides	Potential antibacterial agents.	—	18
Chalcone penicillanate	2-Fluoro-3-nitro-4-chalconyl-6-phenylacetamidopenicillanate possesses antibacterial action.	—	19, 20
Various chalcones	Bacteriostatic activity.	*B. abortus*	21
Furan analogues of chalcone	Significant bacteriostatic action.	*E. coli* *S. aureus* *B. mycoidis* *B. subtilis* *S. lutea*	1
Furan and 8-hydroxyquinoline analogue of chalcone	—	*B. subtilis*	22

Antimicrobial Activity

Compound	Remarks	Test Organism	Ref.
α-Substituted chalcones	Activity was increased by α-bromination.	Trichophyton	23
Hydroxycarboxychalcones and their dihydro derivatives	4'-Hydroxy-5'-carboxychalcones, 4-chloro-4'-hydroxy-5'-carboxychalcone, and 4-methoxy-4'-hydroxy-5'-carboxydihydrochalcone possess antimicrobial activity (*in vitro*).	*B. subtilis* *S. hemolyticus* Other bacteria *in vitro*	24
Chalcone derivative	Methylene dithiodiacetic acid derivative (produced	*Trichomonas vaginalis*	25

217

Compound	Remarks	Test Organism	Ref.

BIOLOGICAL ACTIVITIES OF CHALCONES

Compound	Remarks	Test Organism	Ref.
	by the reaction of chalcone with monothiol acetic acid) has antimicrobial action.		
α-Bromochalcones	Antimicrobial activity.	—	26
Antibiotic Activity Chalcone	Antibiotic activity is associated with the C=C bond of the chalcone molecule. Enhancement in antibiotic activity results when substitution is made at the α-(nitro- and bromo-) and β-(bromo- and hydroxyl-) positions. Addition of cystein or serum to chalcones hampers activity owing to their reduction with the SH group.	—	27
Salicylic chalcones	2'-Hydroxy-4'-carbethoxychalcone has an antibiotic activity.	—	28
Furan analogue of chalcone	Exhibits an antibiotic action.	E. coli	29
Tuberculostatic and Antitubercular Activity Furan analogue of chalcone	Compared to thiomicid it possesses lesser tuberculostatic action and is markedly toxic.	Rats	30
Chalcone-2-hydroxy-4-acetamidobenzene sulfonyl hydrazone	Potential antitubercular compound.	—	31
Semicarbazone and thiosemicarbazones of chalcone or its analogue	The highly effective antitubercular compounds are 1. Semicarbazones of	Tubercle bacilli	32 33 34

Compound	Remarks	Test Organism	Ref.

BIOLOGICAL ACTIVITIES OF CHALCONES

| | thiophene analogue of chalcone containing chloro and nitro substituents. 2. Those corresponding to the general formula p-ROC_6H_4CH=CY—$C\phi NNHSCNH_2$ (R = alkyl, Y = MeO, COOH, or phenyl). | | |

| Chalcone derivative | | | 35 |

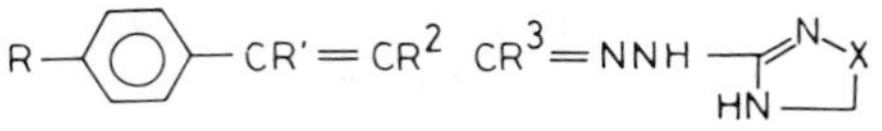

$$R-\langle O \rangle-CR'=CR^2 \quad CR^3=NNH-\overset{N}{\underset{HN}{\diagdown}}X$$

R = Cl; OEt; Ph; CF₃; Br; SMe; SO₂Me; Me, OC(S)NMe₂

R' = H; 4-ClC₆H₄; 2-naphthyl

R² = H; Me

R³ = substituted phenyl; 2-naphthyl

X = (CH₂)₁₋₃; CHMe

Antiparasitic Activity

| Thiophene analogues of chalcone | As possible antiparasitic compounds. | — | 36 |

Antimalarial Activity

| Chalcone derivatives | | | 37 |

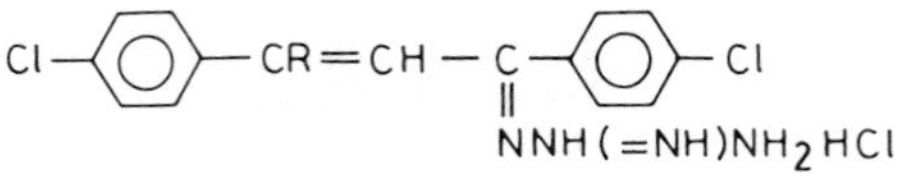

$$Cl-\langle O \rangle-CR=CH-\underset{\underset{NNH(=NH)NH_2 HCl}{\|}}{C}-\langle O \rangle-Cl$$

(R=H; Me; or $\underline{p}$-ClC₆H₄)

Acaricidal Activity

| Chalcone and some furan analogues | Exhibit acaricidal activity. | — | 38 |

Compound	Remarks	Test Organism	Ref.

BIOLOGICAL ACTIVITIES OF CHALCONES

Schistosomicidal Activity

Compound	Remarks	Test Organism	Ref.
Chalcone derivatives	Thienylpyrazolines derived from appropriate chalcones are described as potential schistosomicidal agents.	—	39

Anthelminitic Activity

Compound	Remarks	Test Organism	Ref.
2′,4′-Dihydroxychalcone(s) and derivatives	(a) Toxicity and local irritation are less than hexylresorcinol.	*Ascaris*	40
	(b) Chalcone derivative, 3,5-diphenylisoxazoline, has shown anthelmintic activity.	Pinworms	41
	(c) 2,2′-Dihydroxychalcone has also (*in vitro*) anthelmintic activity.	Amoebae	41
Chalcone derivatives	Methylenedithiodiacetic acid derivative of chalcone possesses antiprotozoal activity.	*Shigella dysenteriae*	25

Insect Repellent and Insecticidal Activity

Compound	Remarks	Test Organism	Ref.
Chalcone derivatives	Some chalcone derivatives are claimed to have insect-repellent properties.	—	42
Chalcone derivatives	*N*-Substituted *o*-carbamoyloxime of chalcone are reported to exhibit a weak insecticidal activity.	—	43
Chalcone	Toxicity toward summer eggs and adult females of fruit tree red spider mite.	*Metatetranychus ulmi*	44
Chalcone α,β-dichloride and DDT	Female flies mortality is 92%.	DDT-resistant female flies	45

Compound	Remarks	Test Organism	Ref.

BIOLOGICAL ACTIVITIES OF CHALCONES

Toxicity to Animals

Compound	Remarks	Test Organism	Ref.
Furan analogue of chalcone	Markedly toxic.	Rats	30
Hydroxy- and methoxy-chalcones	When all the hydroxyl groups (in the poly-hydroxychalcone) are methylated, the toxicity increases.	Freshwater fish	46

Herbicidal Activity

Compound	Remarks	Test Organism	Ref.
Chalcone derivative	(a) N-Substituted o-car-bamoyloxime of chalcone exhibits weak herbicidal ac-tivity.	—	43
	(b) Pyrazolium salts of chalcone are very ef-fective as herbicides.	Lamb's-quarters and mustard	47

Fungistatic and Fungicide Activity

Compound	Remarks	Test Organism	Ref.
Substituted chalcones	2'-Hydroxychalcone sul-fonic acid possesses a weak antifungal activity.	Mold and fungi	48
	—	—	49
	Carboxylic and sulfonic acid derivatives of chal-cone show fungicidal ac-tion.	—	17

(a)

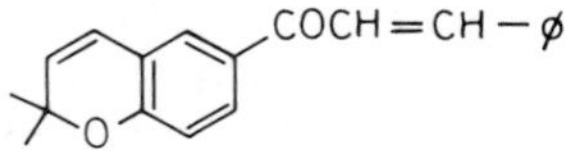

$$\text{COCH}=\text{CH}-\phi$$

(b)

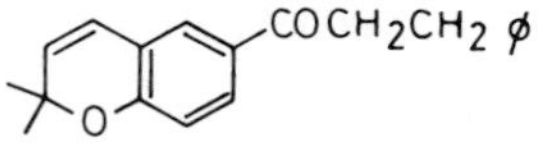

$$\text{COCH}_2\text{CH}_2\ \phi$$

Compound	Remarks	Test Organism	Ref.
Substituted chalcones and dihydrochalcones		*Helminthosporium oryzae*	50

Compound	Remarks	Test Organism	Ref.

BIOLOGICAL ACTIVITIES OF CHALCONES

Compound	Remarks	Test Organism	Ref.
(c) 2',4'-Dihydroxydihydrochalcone.		*Alternaria solani* *Curvularia lunate* (*in vitro*)	
Chalcone	Exhibits an antifungal activity.	Fungi	1, 3
Heterocyclic analogues of chalcone	Furan and 8-Hydroxyquinoline-type compounds, exhibit antifungal activity.	Fungi	1, 22
α-Chalcone and α-2-furan analogue of chalcone	Fungistatic activity is associated with these compounds.	*F. graminearum* *P. digitatum* *B. allii*	51
2-Hydroxy-2'-carboxylic chalcone	Exhibits an antifungal activity.	Cucumber mildew	52

Action on Enzymes

Compound	Remarks	Test Organism	Ref.
3,3',4,4'-Tetrahydroxychalcone	Efficient inhibitor of liver xanthine oxidase activity.	Rats	53, 54
(a) 3,4-Dihydroxy-3'-carboxychalcone (b) 3,4,4'-Trihydroxy-3'-carboxychalcone	Inhibits 5-hydroxytryptophan decarboxylase.	—	55, 56
(a) Naringenin chalcone (b) Hesperidin chalcone (c) Coreopsin (d) Phlorizin (e) Asebotin	Inhibitor of sodium–potassium-dependent ATPase. The extent of inhibition is dependent upon the number and position of the hydroxyl groups. The *para*-hydroxylic function of ring B (phorizin) is involved in the inhibitory action.	Pig kidney	57
Some salicylic chalcones	Inhibits the aromatic L-amino acid carboxylases, (the standard for comparison is α-methyldopa).	Guinea pig kidney	58

Compound	Remarks	Test Organism	Ref.

BIOLOGICAL ACTIVITIES OF CHALCONES

Compound	Remarks	Test Organism	Ref.
Chalcone and furan analogue of chalcone	(a) These compounds inhibit the activity of papain.	Papain	59
	(b) Inhibits cholinesterase in horse serum.	—	55, 56
	(c) Furan analogue of chalcone has a marked ability to inhibit the activity of enzyme dihydroxyphenylalanine decarboxylase.	—	60
Various chalcones and dihydrochalcones	2',4,4'-Trihydroxychalcone is a potent stimulator of indole acetic acid oxidase. It stimulates wheat root growth, inhibits the absorption of sugars and of 2,4-dichlorophenoxyacetic acid, and inhibits oxidative phosphorylation and gives a distinct uncoupling effect.	Wheat root	61
Resorcylic chalcones	Inhibitory activity on L-dopa decarboxylase.	—	62

Antipeptic Ulcer Activity
Isopentenylchalcone

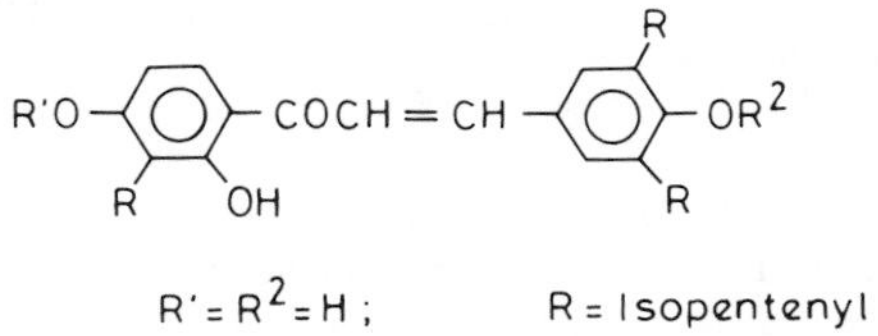

$R'O$—⟨ring⟩—$COCH = CH$—⟨ring⟩—OR^2

R ... OH (on first ring); R ... R (on second ring)

$R' = R^2 = H$; $R = $ Isopentenyl

63–67

These compounds have proved useful in the treatment of rat stomach ulcers.	Rat	

Compound	Remarks	Test Organism	Ref.

BIOLOGICAL ACTIVITIES OF CHALCONES

Compound	Remarks	Test Organism	Ref.
Chalcone derivatives	Useful in the treatment of gastrointestinal ulcers.	Rats	68–76
Prenyloxychalcones	Possess antipeptic ulcer activity.	Rat	77
Prenylchalcones	Exhibit antipeptic ulcer activity.	Mice	78

Hypotensive/Antihypertensive Activity

Compound	Remarks	Test Organism	Ref.
2′,4′,6′-Trihydroxychalcone	Hypotensive property is associated with this chalcone.	—	79
(a) ω-Aminoalkoxy chalcones and acid addition salts (b) N,N-Disubstituted-2-(ω-aminoalkoxy)-3′,4′,5′-trimethoxychalcone (c) Reduced benzofuran chalcone derivative	Compounds with antihypertensive properties.	—	80–84
Indole analogues of chalcone	Weakly hypotensive.	—	85
2-[3-(4-Methyl-1-piperazinyl)propoxy-4′-methyl (and 4′-chloro-)]-chalcone hydrochlorides and related compounds	Hypotensive action.	—	86, 87
4-Aminohalogenochalcone (without a 2′-hydroxy substituent)	Hypotensive activity.	—	49

Antitumor Activity/Cytotoxic Activity

Compound	Remarks	Test Organism	Ref.
2,4,4′-Trihydroxychalcone and 2′,4,4′,6′-tetrahydroxychalcone	Antineoplastic action on Ehrlich's ascitic sarcoma in mice.	Mice	88
Nitrochalcones, having nitro group in 2- (and 2′-) and 4- (and 4′-) positions	These compounds show cytotoxic activity.	Tested against normal and Rous virus-transformed hamster fibroblasts	89

(Continued)

Compound	Remarks	Test Organism	Ref.

BIOLOGICAL ACTIVITIES OF CHALCONES

Isothiocyanatochalcones	4-Nitro-3'-isothiocyana-tochalcone is the most active compound in respect of cytoxicity and cancerostatic effect.	HeLa cells	90
Uvaretin (a dihydrochalcone from *Uvaria acuminata*)	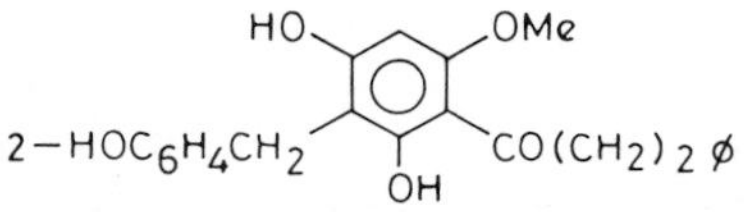		91

$2-HOC_6H_4CH_2$ ... HO ... OMe ... $CO(CH_2)_2\phi$ ... OH

	Uvaretin showed an antitumor activity in a lymphocytic leukemia test.	—	
Flavonol derived from 2'-hydroxy-2,4',5',6,6'-pentamethoxychalcone	Potential antitumor compound.	—	92

Choleretic/Hypocholeretic Activity

(a) 4,4'-Dihydroxy-2'-methoxychalcone	These compounds exhibit a greater activity than the currently used choleretics.	Rats	93
(b) 2',4,4'-Trimethoxy-chalcone (vesidryl)			61
(a) 3'-Nitro-4'-hydroxy-2-methoxy chalcone	Possesse choleretic effect *in vivo*. Compound (a) has the highest choleretic effect.	Rats	6
(b) 3'-Nitro-4'-hydroxy-2,3-dimethoxy-chalcone			
(c) 3'-Nitro-4'-hydroxy-2,5-dimethoxy chalcone			
2',4',6'-Trihydroxychalcone	Hypocholeretic action.	—	79

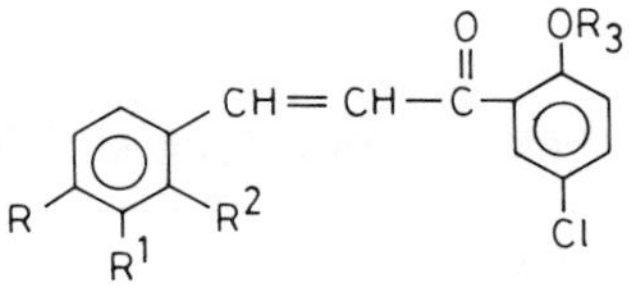

R ... R^1 ... R^2 ... $CH=CH-C$ ... O ... OR_3 ... Cl

a) $R = R^1 = H$; $R^2 = OMe$

225

Compound	Remarks	Test Organism	Ref.

BIOLOGICAL ACTIVITIES OF CHALCONES

Compound	Remarks	Test Organism	Ref.
(a) R_3 = 2-Morphin-oethyl sulfate (b) R = R_1 = R_2 = H; R_2 = CH_2COOH			94
Pyridine analogue of chalcone	Choleretic property.	—	95

Spasmolytic Activity

Compound	Remarks	Test Organism	Ref.
Chalcone; 2 (and 2'-)-Hydroxychalcones; 2,4-dihydroxychalcone; 2,2'-dihydroxychalcone; 2,4,4'-trihydroxy chalcone; and 2,2',4-trihydroxychalcone	Exhibit spasmolytic action.	Isolated bowels and stomach of rats	96
Salts of N-substituted 4'-aminoalkoxy-2',4-dihydroxychalcones	These compounds possess relatively high muscular spasmolytic and low neurospasmolytic properties.	Mouse	97
Reduced benzofuran-chalcone derivatives	Compound with R = 4-OH, and NR^1R^2 = piperidine has the highest spasmolytic activity.	—	83

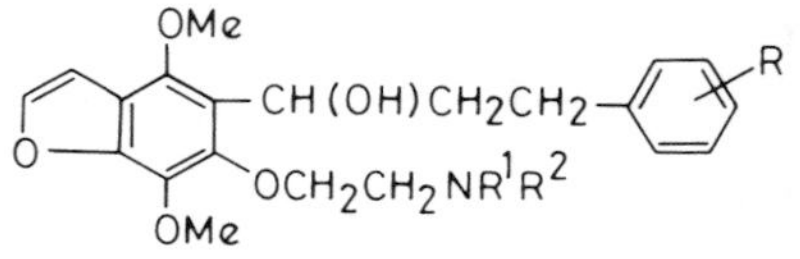

Compound	Remarks	Test Organism	Ref.
Bis (phenylalkyl)amines, the catalytic reduction products of chalcone oxime	Potential spasmolytic compounds.	—	98
Licurzid (chalcone)	Produces spasmolytic action on isolated sections of intestines.	Rat and guinea pig	99

Compound	Remarks	Test Organism	Ref.

BIOLOGICAL ACTIVITIES OF CHALCONES

Antispasmodic and Tranquilizing Action

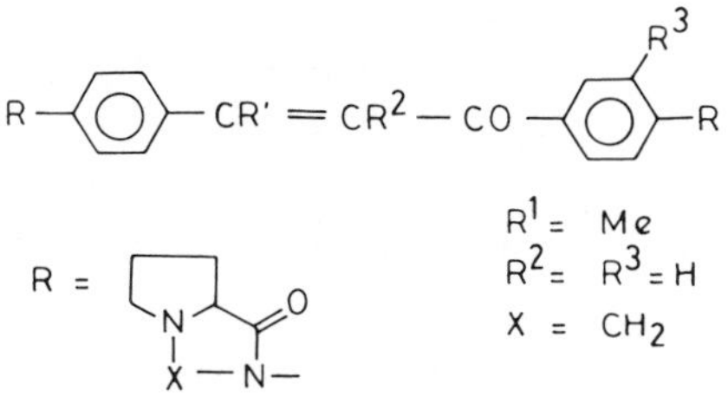

100, 101

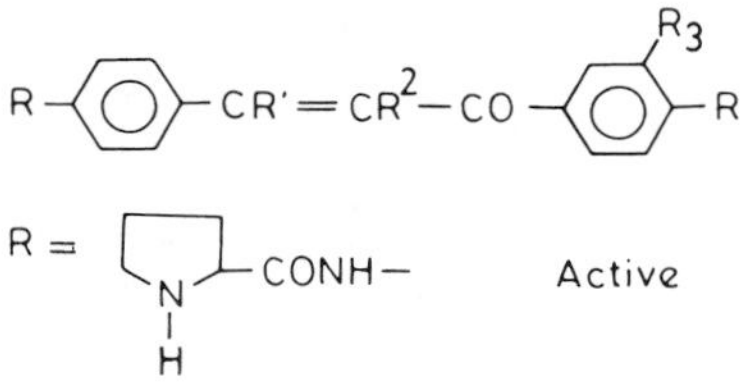

$$R-\bigcirc-CR'=CR^2-CO-\bigcirc^{R^3}-R$$

$R = $ (pyrrolidinone structure)

$R^1 = Me$
$R^2 = R^3 = H$
$X = CH_2$

102

$$R-\bigcirc-CR'=CR^2-CO-\bigcirc^{R_3}-R$$

$R = $ (pyrrolidine-CONH— structure)   Active

i) $R' = R^2 = H$; $R^3 = Me$;
ii) $R' = H$; $R^2 = CH_3$; $R^3 = H$
iii) $R' = CH_3$; $R^2 = R^3 = H$

Antiinflammatory Activity

Substituted chalcones	—	—	49
5-Cinnamoylsalicylic acid	—	—	24

Analgesic and Sedative Action

Substituted chalcones	—	—	49
5-(4-Chlorocinnamoyl) salicylic acid	Analgesic effect is similar to that of aspirin.	—	24

Antithrombic Activity

Substituted chalcones	—	—	49
Flavone derived from the following chalcone:	Potential antithrombic active compound.	—	103

227

Compound	Remarks	Test Organism	Ref.

BIOLOGICAL ACTIVITIES OF CHALCONES

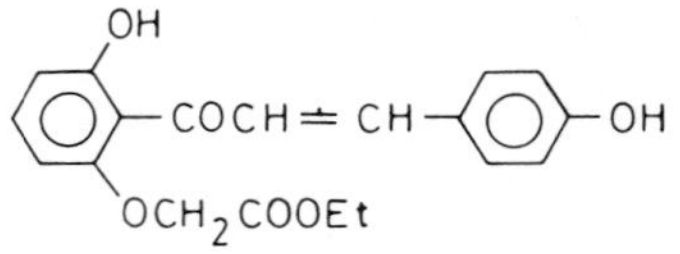

Capillary Fragility

Compound	Remarks	Test Organism	Ref.
Various chalcones	Restore capillary resistance.	Guinea pigs	104
	Decreases capillary fragility and also effects venous circulation.	—	105

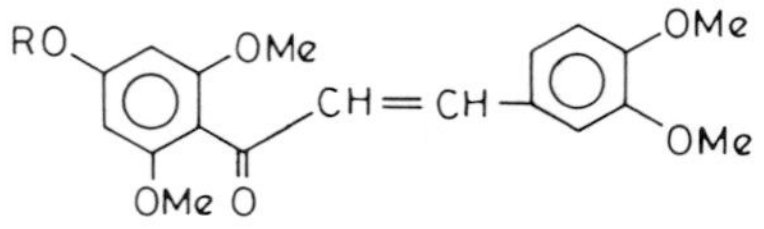

Compound	Remarks	Test Organism	Ref.
Chalcone and hesperidinmethylchalcone	These compounds effect the fragility of capillaries present in the inner surface of the abdominal skin.	Mice	106

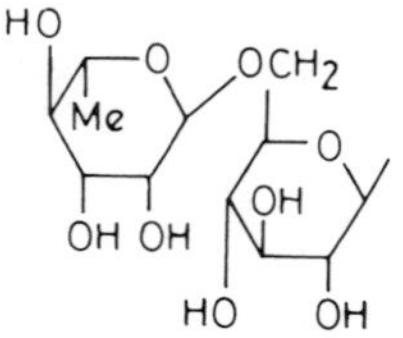

Vasodilatory Activity

Compound	Remarks	Test Organism	Ref.
(a) ω-Aminoalkoxychalcones and acid addition salts (b) Reduced benzofuranchalcone derivatives	Coronary vasodilatory properties are associated with these compounds.	—	82–84, 107

Compound	Remarks	Test Organism	Ref.

BIOLOGICAL ACTIVITIES OF CHALCONES

Compound	Remarks	Test Organism	Ref.
Mecinarone (a benzofuranic chalcone)	Vasodilatory activity (on the peripheral and cerebral circulation).	Experimental animals	108
Pyridine analogues of chalcone	Coronary vasodilative properties.	—	95

Estrogenic Activity

Compound	Remarks	Test Organism	Ref.
2'-Hydroxy (and 2'-chloro-)-3,4-methylenedioxy-4'-fluorochalcones, and 2'-chloro-4,4'-difluorochalcone	These compounds produced uterotropic effect, decreased the weight and size of the testes and seminal vesicles and inhibited implantation in the mouse.	Mice	109
4,2',4'-Trimethoxchalcone epoxide, and 4-methoxy-2',4'-dibenzoyloxychalcone epoxide	Estrogenic activity (?)	Rats	110

Anesthetic Activity

Compound	Remarks	Test Organism	Ref.
Pyrazole derived from 4-dimethylamino chalcone	It has some local anesthetic activity.	—	111

Anticoagulating Effect

Compound	Remarks	Test Organism	Ref.
2',4',6'-Trihydroxychalcone	Anticoagulating properties.	—	79

Anticonvulsant/Narcotic Potentiation Activity

Compound	Remarks	Test Organism	Ref.
3,4-Methylenedioxychalcone	It has proved effective as anticonvulsant; also shows narcotic potentiation activity.	—	112

Therapeutic Activity

Compound	Remarks	Test Organism	Ref.
Cyanomethylchalcones	These compounds are claimed as valuable medicinal agents for cardiovascular diseases and endocrine dysfunctions.	—	113

229

Compound	Remarks	Test Organism	Ref.

BIOLOGICAL ACTIVITIES OF CHALCONES

Compound	Remarks	Test Organism	Ref.
Hesperidin methylcarboxychalcone	It exerts therapeutic action in the treatment of chronic diseases of the eye and kidney, including rheumatoid diseases such as bursitis and osteoarthritis.	—	114
Hesperidin methylchalcone	The compound, when incorporated in the diet (0.2%), shows an inhibitory effect on the incidence of dental caries.	Cotton rats	115

Antiangiotensin/Antiarrhythmic/Diuretic Activity

Compound	Remarks	Test Organism	Ref.
Reduced benzofuranchalcone derivative (cf. spasmolytic activity)	These compounds possess antiangiotensin, antiarrhythmic, and diuretic activity.	—	85

Miscellaneous Biological Activities

Compound	Remarks	Test Organism	Ref.
Chalcone, 2-hydroxychalcone, 2',3-dihydroxychalcone, and hesperidinchalcone	These compounds are able to protect epinephrine from destruction (*in vitro*).	Isolated strip of intestine	116
Dialkylaminoalkoxy derivative of chalcone	Potential adrenergic blocking agents.	—	117
Aminoazachalcones	Adrenal cortex inhibitors.	Rats	118
Azachalcones:	Suprarenal gland inhibitors of the amphenone type. However, these compounds are less active than metyrapon.	Rat	122

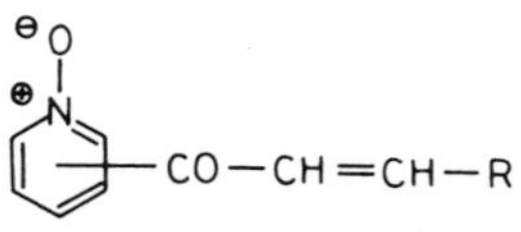

R = 2-, 3-, 4-pyridyl;
R = Ph; 4-NH(Me)C_6H_4
or
R = 4-N(Me)$_2C_6H_4$

Compound	Remarks	Test Organism	Ref.

BIOLOGICAL ACTIVITIES OF CHALCONES

Compound	Remarks	Test Organism	Ref.
Substituted chalcones	The most potent compound in pharmacological activity is 2-(2-dimethylaminoethoxy) chalcone, but it does not compare with the available therapeutic agents in specificity, potency, and duration of action.	—	119
Licurzid (chalcone)	(a) Reduces stomach motility. (b) Inhibits evacuation of water from stomach to deudenum. (c) Inhibits development of exudative processes in the inflammation and prevents development of neurogenic and butadione stomach ulcers.	Rats and mice	99
2-[2-Dimethylaminoethoxy)-3′,4′,5′-trimethoxy] chalcone hydrochloride	(a) This compound is an effective and long active depressor agent. (b) There is an electrolytic alteration between blood vascular smooth muscles following treatment with the chalcone.	Dogs and rats	120
Sulfur-containing derivative of chalcone, $\phi COCH_2S$–C_6H_4Cl, obtained by the interaction of the chalcone epoxide with thiol	Biological activity.	—	121

(Continued)

Compound	Remarks	Test Organism	Ref.

BIOLOGICAL ACTIVITIES OF CHALCONES

Action on Spermatocytic Chromosomes

Compound	Remarks	Test Organism	Ref.
2′,4′-Dihydroxychalcone	The rejoining of broken ends of chromosomes and chromatids is accelerated by the addition of chalcone.	Grasshopper	123

Decrease in the Incidence of Blood Spots in Chicken Eggs

Compound	Remarks	Test Organism	Ref.
(a) 3-Pyrrole-2-aldehyde chalcone	Active.	Chicken eggs	124
		Chicken eggs	125
(b)			

Plant Growth-Inhibiting Activity

Compound	Remarks	Test Organism	Ref.
3,2′,4′-Trihydroxy-4-methoxydihydrochalcone	Inhibitory effect on the growth of plant triticale.	Triticale	126

REFERENCES

1 Geiger, W. B., and Conn, J. E., *J. Am. Chem. Soc.*, **67**, 112 (1945).

2 Schraufstätter, E., *Experientia*, **4**, 484 (1948).

3 Ishida, S., Matsuda, A., Kawamura, Y., and Yamanaka, K., *Chemotherapy (Tokyo)*, **8**, 146 (1960); *Chem. Abstr.*, **54**, 22844c (1960).

4 Goncalves de Lima, O., Francisco de Mello, J., Coelho, J. S. de B., Lyra, F. D. de A., Fernandes de Albuquerque, M. N., Marini-Bettolo, G. B., Delle Monache, G., and Delle Monache, F., *Farmaco, Ed. Sci.*, **30**, 326 (1975); *Chem. Abstro.*, **83**, 91495t (1975).

5 Gabor, M., Sallai, J., Széll, T., and Sipos, G., *Acta Microbiol. Acad. Sci. Hung.*,**14**, 45 (1967); *Chem. Abstr.*, **67**, 51338g (1967).

6 Gabor, M., Sallai, J., and Széll, T., *Kiserl. Orvostud.*, **20**, 540 1968); *Chem. Abstr.*, **70**, 56156t (1969).

7 Gabor, M., Sallai, J., Széll, T., *Acta Physiol. Budapest*, **34**, 227 (1968); *Chem. Abstr.*, **71**, 11692u (1969).

8 Gabor, M., Sallai, J., and Széll, T., *Arch. Pharm. (Weinheim)*, **303**, 593 (1970); *Chem. Abstr.*, **73**, 106605y (1970).

9   Marrian, D. H., Russell, P. B., and Todd, A. R., *J. Chem. Soc.*, 1419 (1947).

10 Ambekar, S., Vernekar, S. S., Acharya, S., and Rajagopal, S., *J. Pharm. Pharmacol.*, **13**, 698 (1961).

11 Gudi, N., Hiremath, S., Badiger, V., and Rajagopal, S., *Arch. Pharm.*, **295**, 16 (1962); *Chem. Abstr.*, **57**, 7154e (1962).

12 Jerchel, D., and Oberheiden, H., *Angew. Chem.*, **67**, 145 (1955).

13 Schraufstätter, E., and Deutsch, S., *Z. Naturforsch.*, **3b**, 430 (1948).

14 Schraufstätter, E., and Deutsch, S., *Z. Naturforsch.*, **3b**, 163 (1948).

15 Covello, M., Dini, A., and Piscopo, E., *Rend. Accad. Sci. Fis. Mat., Naples*, **37**, 56 (1970); *Chem. Abstr.*, **76**, 33900r (1972).

16 Hirose, K., Vkai, S., and Hattori, T., *Yakugaku Zasshi*, **91**, 604 (1971); *Chem. Abstr.*, **75**, 129467k (1971)

17 Wurm, G., *Arch. Phar. (Weinheim)*, **308**, 142 (1975).

18 Calcinari, R., *Farmaco, Ed., Sci.*, **27**, 397 (1972); *Chem. Abstr.*, **77**, 48007z (1972).

19 Aries, R., German Patent, 2,341,514 (1974); *Chem. Abstr.*, **80**, 146152z (1974).

20 Societe de Recherches, French Patent, 2,181,505; *Chem. Abstr.*, **80**, 108509p (1974).

21 Jeney, E., and Zsolnai, T., *Acta Microbiol. Acad. Sci. Hung.*, **2**, 249 (1955); *Chem. Abstr.*, **49**, 13501g (1955).

22 Kushwaha, S. C., Dinkar, and Lal, J. B., *Indian J. Chem.*, **5**, 82 (1967).

23 Gasha, M., Tsjuji, A., Sakurai, Y., Kurumi, M., Endo, T., Sato, S., and Yamaguchi, K., *Yakugaku Zasshi*, **92**, 719 (1972); *Chem. Abstr.*, **77**, 97221y (1972).

24 Lespagnol, A., Lespagnol, C., Lesieur, D., Cazin, J. C., Cazin, M., Beerens, H., and Romond, C., *Chim. Ther.*, **7**, 365 (1972); *Chem. Abstr.*, **78**, 52753c (1973).

25 Kishimoto, Y., Akabori, Y., and Horiguchi, T., *Yakugaku Zasshi*, **78**, 447 (1958); *Chem. Abstr.*, **52**, 13863b (1958).

26 Yamaguchi, K., Sakurai, Y., and Kurumi, M., Japanese Patent, 72 47,016 (1972); *Chem. Abstr.*, **78**, 97330d (1973).

27 Schraufstätter, E., and Deutsch, S., *Z. Naturforsch.*, **4b**, 276 (1949).

28 Lespagnol, A., Lespagnol, C., Lesieur, D., Bonte, J. P., Blain, Y., and Labiau, O., *Chim. Ther.*, **6**, 192 (1971); *Chem. Abstr.*, **75**, 98285t (1971).

29 Kamoda, M., and Ito, N., *J. Agr. Chem. Soc. Japan*, **28**, 799 (1954); *Chem. Abstr.*, **49**, 5573b (1955).

30 Jeney, E., Zsolnai, T., and Lázár, J., *Zentr. Bakteriol. Parasitenk, Abt. I, Orig.*, **163**, 291 (1955); *Chem. Abstr.*, **49**, 15074g (1955).

31 Lora-Tamayo, M., Mūnicio, A. M., and Ruiz, J. L., *An. R. Soc. Esp. Fiz. Quim. (Madrid)*, **56B**, 403 (1960); *Chem. Abstr.*, **55**, 7410a (1961).

32 Buu-Hoi, N. P., Xuong, N. D., and Sy, M., *Bull. Soc. Chim. France*, 1646 (1956); *Chem. Abstr.*, **51**, 12063i (1957).

33 Bayer, F., British Patent 708,013 (1954); *Chem. Abstr.*, **50**, 1086b (1956).

34 Schmidt, H., Behnisch, R., and Schraufstätter, E., U.S. Patent 2,676,978 (1954); *Chem. Abstr.*, **49**, 7596h (1955).

35   Tomcufcik, A. S., Wilkinson, R. G., and Child, R. G., German Patent 2,502,490 (1975); *Chem. Abstr.*, **83**, 179067r (1975).*

36   Laliberate, R., Manson, J., Warwick, H., and Medawar, G., *Can. J. Chem.*, **46**, 1952 (1968).

37   Henry, D. W., U.S. Patent, 3,726,920 (1973); *Chem. Abstr.*, **79**, 18450n (1973).

38   Velarde, E., *Bol. Soc. Quim. Peru*, **36**, 127 (1970); *Chem. Abstr.*, **75**, 19808q (1971).

39   El-Kerdawy, M. M., Samour, A. A., and El-Agamey, A. A., *Pharmazie*, **30**, 76 (1975); *Chem. Abstr.*, **83**, 9888s (1975).

40   Takayanagi, Y., *Ann. Rep. Tohoku Coll. Pharm. No. 1*, 10 (1954); *Chem. Abstr.*, **50**, 4389e (1956).

41   Laliberté, R., Campbell, D., and Bruderlein, F., *Can. J. Pharm. Sci.*, **2**, 37 (1967); *Chem. Abstr.*, **67**, 98058f (1967).

42   Nakanishi, M., and Tsuda, A., Japanese Patent, 74 46,056; *Chem. Abstr.*, **83**, 38818z (1975).

43   Ruchkin, V. E., Shvetsova-Shilovskaya, K. D., and Melńikov, N. N., *Probl. Poluch. Poluprod. Prom. Org. Sin., Akad. Nauk. SSSR, Otd. Obshch. Tekh. Khim.*, **66** (1967); *Chem. Abstr.*, **68**, 95417g (1968).

44   Eaton, J. K., and Davies, R. G., *Ann. Applied Biol.*, **37**, 471 (1950); *Chem. Abstr.*, **45**, 3111h (1951).

45   Clark, S. F., U.S. Patent, 2,805,184; *Chem. Abstr.*, **52**, P 2325i (1958).

46   Narasimhachari, N., and Seshadri, T. R., *Proc. Indian Acad. Sci.*, **27A**, 128 (1948).

47   Cross, B., U.S. Patent 3,925,408 (1975); *Chem. Abstr.*, **84**, 135646v (1975).

48   Wurm, G., and Lachmann, C., *Arch. Pharm.*, **307**, 695 (1974); *Chem. Abstr.*, **81**, 169399v (1974).

49   Hsu, K. K., and Chen, F. C., *Tai-wan K'o Hsueh*, **27**, 23 (1973); *Chem. Abstr.*, **80**, 66597h (1974).

50   Chowdhury, A., Mukherjee, N., and Adityachaudhury, N., *Experientia*, **30**, 1022 (1974).

51   McGowan, J. C., Brian, P. W., and Hemming, H. G., *Ann. Applied Biol.*, **35**, 25 (1948); *Chem. Abstr.*, **43**, 9334b (1949).

52   Ko, K., Iseki, K., Adachi, Y., Konno, K., Yokoyoma, K., and Misato, T., *Nippon Nayaku Brakkaishi*, **3**, 155 (1978); *Chem. Abstr.*, **89**, 192310h (1978).

53   Martin, G. J., Beiler, J. M., and Avakian, S., U.S. Patent 2,769,817 (1956); *Chem. Abstr.*, **51**, 14815d (1957).

54   Beiler, J. M., Graff, M. and Martin, G. J., *Am. J. Digestive Diseases*, **19**, 333 (1952); *Chem. Abstr.*, **47**, 1198e (1953).

55   Kamoda, M., Chiba, T., Mori, K., and Ito, N., *Botyu Kagaku*, **18**, 117 (1953); *Chem. Abstr.*, **48**, 778c (1954).

56   Yuwiler, A., Geller, E., and Eiduson, S., *Arch. Biochem. Biophys.*, **89**, 143 (1960); *Chem. Abstr.*, **54**, 21222c (1960).

57   Hase, J., Kobashi, K., and Kobayashi, R., *Chem. Pharm. Bull.*, **21**, 1076 (1973); *Chem. Abstr.*, **80**, 10261b (1974).

58 Lesieur-Demarquilly, I., Mizon, J., and Osteux, R., *Ann. Pharm. Franc.*, **31**, 705 (1973); *Chem. Abstr.*, **81**, 59921b (1974).

59 Watanabe, M., Okada, K., Mori, K., and Ito, N., *Botyu Kagaku*, **17**, 1 (1952); *Chem. Abstr.*, **46**, 7151a (1952).

60 Deutsch, D. H., and Garcia, E. N., U.S. Patent 2,754,299 (1956); *Chem. Abstr.*, **51**, 4437h (1957).

61 Stenlid, G., *Physiol. Plant.*, **21**, 882 (1968); *Chem. Abstr.*, **69**, 84194b (1968).

62 Lesieur-Demarquilly, I., and Lesieur, D., *Bull. Soc. Pharm.*, **34**, 9 (1978).

63 Komatsu, M., Kyogoku, K., Hatayama, K., Suzuki, K., and Yukomori, S., Japanese Patent 74 05,950 (1974); *Chem. Abstr.*, **80**, 120551d (1974).

64 Kyogoku, K., Hatayama, K., Yokomori, S., and Seki, T., Japanese Patent 74 126,654 (1974); *Chem. Abstr.*, **83**, 78864t (1975).

65 Kyogoku, K., Hatayama, K., Yokomori, S., and Seki, T., Japanese Patent 75 24,258 (1975); *Chem. Abstr.*, **83**, 178567 (1975).

66 Kyogoku, K., Hatayama, K., Yokomori, S., and Seki, T., Japanese Patent 75 24,257 (1975); *Chem. Abstr.*, **83**, 178567 (1975).

67 Kyogoku, K., Hatayama, K., Yokomori, S., and Seki, T., Japanese Patent 75 140,430 (1975); *Chem. Abstr.*, **84**, 121467e (1975).

68 Kyogoku, K., Hatayama, K., Yokomori, S., Swada, J., and Tanaka, I., Japan Kokai, 77 97, 950 (1977); *Chem. Abstr.*, **88**, 74200d (1978).

69 Vanstone, A. E., and Maile, G. K., Ger. Offen 2,810,253 (1977); *Chem. Abstr.*, **90**, 22628w (1979).

70 Kyogoku, K., Hatakeyama, M., Yokomori, S., Swada, J., and Tanaka, I., Japan Kokai Tokkyo Koho 78 116,355 (1978); *Chem. Abstr.*, **90**, 87056c (1979).

71 Kyogoku, K., Hatakeyama, M., Yokomori, S., Miyata, Y., Swada, J., and Tanaka, I., Japan Kokai Tokkyo Koho 79 19,948 (1979); *Chem. Abstr.*, **91**, 20115y (1979).

72 Biorex Laboratories Ltd., Fr. Demande 2,383,157 (1978); *Chem. Abstr.*, **91**, 39136c (1979).

73 Kyogoku, K., Hatakeyama, M., Yokomori, S., Miyata, Y., Sawada, J., and Tanaka, I., Japan Kokai Tokkyo Koho 79 19,947 (1979); *Chem. Abstr.*, **91**, 20114x (1979).

74 Kyogoku, K., Hatayama, K., Yokomori, S., Sawada, J., and Tanaka, I., Fr. Demande 2,340,924 (1977); *Chem. Abstr.*, **88**, 104910e (1978).

75 Kyogoku, K., Hatakeyama, M., Yokomori, S., Miyata, Y., Sawada, J., and Tanaka, I., Japan Kokai Tokkyo Koho 79 19,949 (1979); *Chem. Abstr.*, **90**, 203699r (1979).

76 Kyogoku, K., Hatakeyama, M., Yokomori, S., Miyata, Y., Sawada, J., and Tanaka, I., Japan Kokai Tokkyo Koho 79 22,349 (1979); *Chem. Abstr.*, **90**, 203690f (1979).

77 Kyogoku, K., Hatayama, K., Yokomori, S., and Seki, T., Belgium Patent 816,463 (1974); *Chem. Abstr.*, **83**, 43051y (1975).

78 Kyogoku, K., Hatayama, K., Yokomori, S., and Seki, T., Japanese Patent 75 140,429 (1975); *Chem. Abstr.*, **85**, 5913g (1976).

79 Lafon, L., Ger. Offen, 2,010,180 (1970); *Chem. Abstr.*, **73**, 120342s (1970).

80 Mizuta, Y., Igasa, H., and Uno, J., and Tsukamoto, M., Japanese Patent 74 21,125 (1974); *Chem. Abstr.*, **82**, 86208b (1975).

81 Igasa, H., Tsukamoto, M., and Uno, J., Japanese Patent 73 37,027 (1973); *Chem. Abstr.*, **80**, 120553f (1974).

82 Igasa, H., Mizuta, Y., Tsukamoto, M., and Uno, J., Japanese Patent 74 21,126 (1974); *Chem. Abstr.*, **82**, 97841d (1975).

83 Raynaud, G., Pourrias, B., Thomas, J., Thomas, M., Roblin, B., Hecaen, V., and Perroud, F., *Eur. J. Med. Chem. Chim. Ther.*, **9**, 85 (1974); *Chem. Abstr.*, **81**, 99240k (1974).

84 Raynaud, G., Pourrias, B., Thomas, J., and Thomas, M., *Chim. Ther.*, **8**, 479 (1973); *Chem. Abstr.*, **81**, 45196g (1974).

85 Venturella, P., Bellino, A., and Piozzi, F., *Farmco, Ed., Sci.*, **26**, 591 (1971); *Chem. Abstr.*, **75**, 88432z (1971).

86 Kinugasa, H., Tsukamoto, M., Mizuta, Y., and Uno, J., Japanese Patent 74 06,904 (1974); *Chem. Abstr.*, **80**. 145797b (1974).

87 Mizuta, Y., Kinugasa, H., Uno, J., and Tsukamoto, M., Japanese Patent 74 06,903 (1974); *Chem. Abstr.*, **80**, 146198u (1974).

88 Kabiev, O. K., and Vermenichev, S. M., *Izv. Akad. Nauk. Kaz. SSR. Ser. Biol.*, **9**, 72 (1971); *Chem. Abstr.*, **75**, 47091u (1971).

89 Dore, J. C., and Viel, C., *J. Pharm. Belg.*, **29**, 341 (1974); *Chem. Abstr.*, **83**, 90650c (1975).

90 Horakova, K., Drobinca, L., and Nemec, P., *Neoplasma*, **18**, 355 (1971); *Chem. Abstr.*, **75**, 107958u (1971).

91 Cole, J. R., Torrance, S. J., Wiedhopf, R. M., Arora, S. K., and Bates, R. B., *J. Org. Chem.*, **41,** 1852 (1976).

92 Uong, T., Yan, C-F., and Chen. F. C., *Hua Hsueh* (4), 133 (1973); *Chem. Abstr.*, **81**, 120380t (1974).

93 Chabannes, B., Gradgeorge, M., Duperray, B., and Pacheco, H., *Chim. Ther.*, **8**, 621 (1973); *Chem. Abstr.*, **81**, 20752x (1974).

94 Flammang, M., Wermuth, C. G., and Delassue, H., *Chim. Ther.*, **5**, 431 (1970); *Chem. Abstr.*, **74**, 125043a (1971).

95 Laboratoires Gerda Fr. M., 6427; *Chem. Abstr.*, **75**, 40406s (1971).

96 Formanek, K., Höller, H., Janisch, H., and Kowac, W., *Pharm. Acta Helv.*, **33**, 437 (1958); *Chem. Abstr.*, **53**, 15380h (1959).

97 Woelm, F. M., German Patent 1,174,311 (1964); *Chem. Abstr.*, **61**, P11933g (1964).

98 Dornow, A., and Frese, A., *Arch. Pharm.*, **285**, 463 (1952); *Chem. Abstr.*, **48**, 11374a (1954).

99 Obolentseva, G. V., and Khadzhai, Y. I., *Vopr. Izuch. Ispol'z. Solodki Nauk.*, 163 (1966); *Chem. Abstr.*, **67**, 107314r (1967).

100 Oshiro, S., Nagura, T., Sugihara, Y., Okamoto, K., Ishida, R., and Shintomi, K., Japanese Patent 73 19,569 (1973); *Chem. Abstr.*, **78**, 147788g (1973).

101 Oshiro, S., Nagura, T., Sugihara, Y., Okamoto, K., Ishida, R., and Shintomi, K., Japanese Patent 73 19,570 (1973); *Chem. Abstr.*, **78**, 147790b (1973).

102 Oshiro, S., Nagura, T., Sugihara, Y., Okamoto, K., Ishida, R., and Shintomi, K., Japanese Patent 73 19,594 (1973); *Chem. Abstr.*, **78**, 147963k (1973).

103 Yan, C. F., Uong, T., and Chen, F-C, *Hua Hsueh* (4), 135 (1973); *Chem. Abstr.*, **81**, 120382v (1974).

104 Huges, E. G., and Parkes, M. W., *Jubilee Vol. Emil, Barell*, 216 (1946); *Chem. Abstr.*, **41**, 2141a (1947).

105 Fabre, P. S. A., French Patent, 2,183,612 (1974); *Chem. Abstr.*, **80**, 121294j (1974).

106 Lockett, M. F., and Jarman, D. A., *Brit. J. Pharmacol.*, **13**, 11 (1958).

107 Uno, J., Kinugasa, H., Tsukamoto, M., and Mizuta, Y., Japanese Patent 72 49,581 (1972); *Chem. Abstr.*, **78**, 147569m (1973).

108 Pourrias, B., Sergant, M., Thomas, J., Gouret, C., and Raynaud, G., *Arzneim.-Forsch.*, **25**, 782 (1975); *Chem. Abstr.*, **83**, 91023n (1975).

109 Jacob, D., and Kaul, D. K., *Acta Endocrinol. (Copenhagen)*, **74**, 371 (1973); *Chem. Abstr.*, **80**, 22992d (1974).

110 Sharma, R. C., Gupta, S. K., Gupta, L., and Arora, R., *Indian J. Exp. Biol.*, **10**, 455 (1972).

111 Hosni, G., and Saad, S. F., *Acta Chim. Acad. Sci. Hung.*, **86**, 263 (1975); *Chem. Abstr.*, **84**, 30959w (1976).

112 Unicler, S. A., French Patent, 2,253,503 (1975); *Chem. Abstr.*, **84**, 89825c (1976).

113 Rorig, K. J., U.S. Patent, 2,755,299 (1956); *Chem. Abstr.*, **51**, 2868c (1957).

114 Hart, B. F., U.S. Patent, 2,962,162 (1960); *Chem. Abstr.*, **54**, 12161g (1960).

115 Thompson, D. T., Vogel, J. J., and Phillips, P. H., *J. Dental. Res.*, **44**, 596 (1965); *Chem. Abstr.*, **63**, 7526d (1965).

116 Wilson, R. H., and DeEds, F., *J. Pharmacol. Exp. Therap.*, **95**, 399 (1949); *Chem. Abstr.*, **43**, 5542a (1949).

117 Packman, A. M., and Rublin, N., *Am. J. Pharm.*, **134**, 45 (1962).

118 Nagy, A., Misikova, E., and Heger, J., *Cesk. Farm.*, **26**, 140 (1977); *Chem. Abstr.*, **88**, 6671h (1978).

119 Rossi, G. C., and Packman, E. W., *J. Am. Pharm. Assoc.*, **47**, 640 (1958).

120 Sherman, G. P., *Diss. Abstr.*, **29B**, 2145 (1968).

121 Ukai, S., Hirose, K., Hattori, T., Kayano, M., and Yamamoto, C., *Yakugaku Zasshi*, **92**, 278 (1972); *Chem. Abstr.*, **77**, 34207c (1972).

122 Durinda, J., Szucs, L., Struharova, L., Kolena, J., and Heger, J., *Cesk. Farm.*, **21**, 276 (1972); *Chem. Abstr.*, **78**, 29576p (1973).

123 Saha, A. K., and Khudabaksh, A. R., *Cytologia*, **39**, 359 (1974); *Chem. Abstr.*, **81**, 146348b (1974).

124 Bigland, C. H., Bennett, E. B., and Abbott, U. K., *Proc. Soc. Exp. Biol. Med.*, **116**, 1122 (1964).

125 Bigland, C. H., Bennett, E. B., and Abbott, U. K., *Poultry Sci.*, **44**, 140 1965).

126 Telenyl, P., *Talajtan*, **26**, 391 (1977); *Chem. Abstr.*, **88**, 164842u (1978).

Chapter Twenty-Seven
Chalcone Epoxides

SYNTHESIS BY DARZEN'S CONDENSATION AND OXIDATION

Epoxy derivatives of chalcones and their heterocyclic analogues, thienyl, selenienyl, and pyridyl, have been prepared by the following methods[1-3]:

1 By Darzen's condensation of aldehydes with haloacyl compounds.
2 By the oxidation of chalcones and their heterocyclic analogues with alkaline hydrogen peroxide. This method has been exploited for the synthesis of the epoxides derived from vinyl-[2] and *o*-methoxychalcones.[3]

REACTIONS

A review on the reactions of substituted chalcone epoxides has been published.[4]

Toluene

Friedel–Crafts alkylation of epoxychalcones with toluene in the presence of anhydrous $AlCl_3$ is reported to yield indenes (**I**).[5]

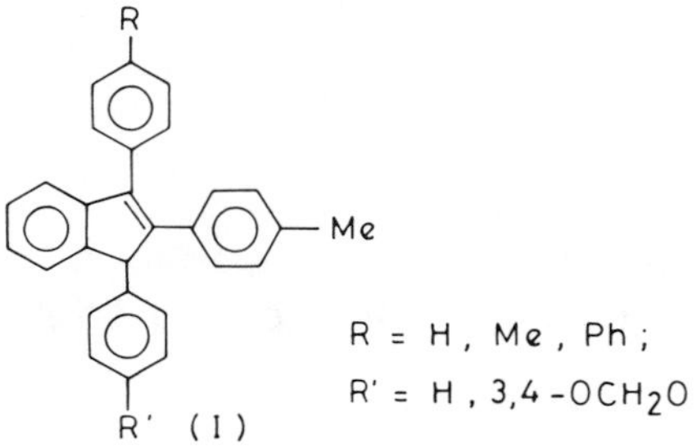

R = H, Me, Ph;
R' = H, 3,4 -OCH₂O

Thiophenols and Thioalcohols[7]

The following example illustrates the types of products (**IV, V, and VII**) formed in the reaction of chalcone epoxide with thiophenol. An analogous reaction occurs with the thioalcohols, $PhCH_2SH$.

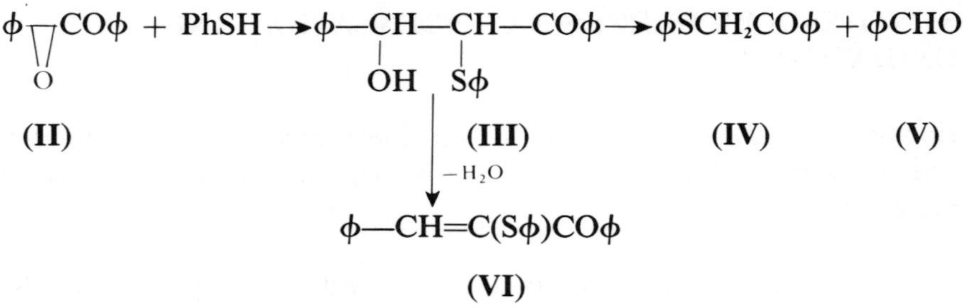

$$\phi\underset{O}{\overset{}{\diamond}}CO\phi + PhSH \longrightarrow \phi-\underset{OH}{\overset{}{CH}}-\underset{S\phi}{\overset{}{CH}}-CO\phi \longrightarrow \phi SCH_2CO\phi + \phi CHO$$

$$\textbf{(II)} \qquad\qquad\qquad\qquad \textbf{(III)} \qquad\qquad \textbf{(IV)} \qquad\qquad \textbf{(V)}$$

$$\downarrow {\scriptstyle -H_2O}$$

$$\phi-CH{=}C(S\phi)CO\phi$$

$$\textbf{(VI)}$$

$$\phi SCH_2CO\phi + \textbf{(VI)} \xrightarrow{\text{Michael condensation}} \phi CH[CH(S\phi)CO\phi]_2$$

$$\textbf{(IV)} \qquad\qquad\qquad\qquad\qquad\qquad\qquad\qquad \textbf{(VII)}$$

Amines[5,8,9]

Several amines, R^2H, react with chalcone epoxides to yield R—$C_6H_4COCH(OH)CHR^2$—C_6H_4R'. The amines used in the reaction are pyridine, morpholine, N-methylaniline, diethylamine, benzylamine, and p-tolylamine.

Hydroxylamines [5,6]

Isoxazoles (VIII) are produced by the interaction of chalcone epoxides and hydroxylamine:

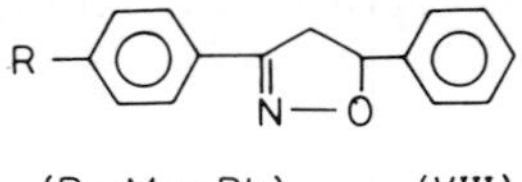

$$(R = Me; Ph) \qquad (VIII)$$

Hydrazines

Substituted chalcone epoxides are reported to react with hydrazine[5,6,9-11] or its substituted derivatives to yield 3,5-diaryl-4-hydroxy-Δ^2-pyrazolines (IX):

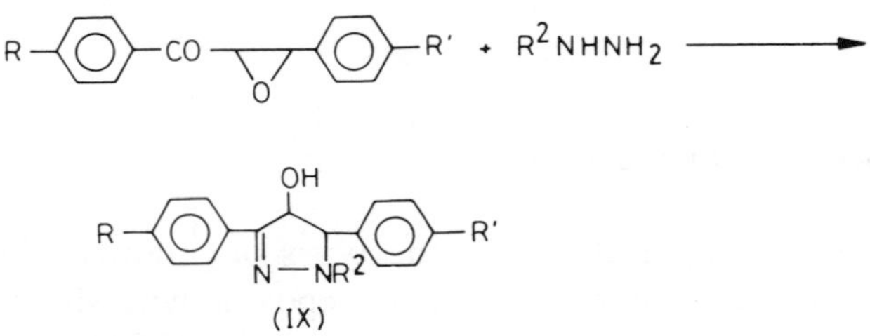

$$(IX)$$

$$R = Me, H; \quad R' = H, MeO; R^2 = H, \phi, C_6H_3(NO_2)_2$$

o-Phenylenediamine

2-Phenylbenzimidazole (10%) has been secured by the reaction of o-phenylenediamine with *trans*-chalcone epoxide.[12]

Reducing Agents

trans-Chalcone epoxide on reduction with lithium aluminum hydride[13] gives the mixture of *erythro*-1,3-diphenyl-1,2-propandiol and ($\pm$)-1,3-diphenyl-1,2-propandiol. The same products are produced when *trans*-chalcone epoxide is catalytically hydrogenated over PtO_2.[13] 2'-Benzyloxychalcone epoxide on treatment with KBH_4 or $LiAlH_4$–$AlCl_3$ (1:7) is reported[14] to yield *trans*-2,3-*cis*-3,5-flavandiol (**X**). However, the epoxide of 2'-methoxymethoxychalcone under these conditions gives a mixture of two isomeric diols, **X** and **XI**, the former predominating:

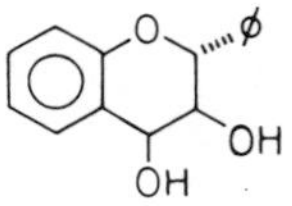

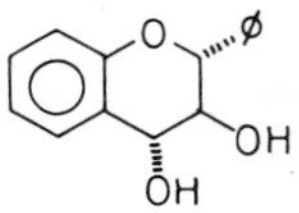

Chalcone and 1,4-dibenzoyl-2,3-diphenylbutane are produced when chalcone epoxide is reduced by chromous chloride.[15]

Acids

With HCl The kinetics and mechanism of the cyclization of 2'-hydroxychalcone epoxide to the 3-hydroxyflavanone in water have been investigated.[16] The treatment of 2'-(4-methoxybenzyl)-4,4'-dimethoxychalcone epoxide with HCl in acetic acid is reported to give *trans*-3-hydroxy-7,4'-dimethoxyflavanone.[17] On the other hand, *trans*-4'-methoxychalcone epoxide on treatment with HCl furnishes the corresponding chlorohydrins.[18] The chlorohydrins have been found by NMR spectroscopy to correspond to *threo* and *erythro* configurations.[18] Under the same conditions 3-nitrochalcone epoxide yields only one isomer, *erythro*.[18] Several chlorohydrins (*erythro* and *threo*) have been synthesized.[19,20] The reaction of hydrazine with several chlorohydrins has been studied.[10]

With Dry HCl in Ether The treatment of 4′-methylchalcone epoxide with dry HCl in ethereal solution yields 1-(p-tolyl)-2-hydroxy-3-chloro-3-phenylpropan-1-one.[21] The effect of various functional groups (R and R′) on the oxirane ring opening by HCl or BF_3 has been studied.[22]

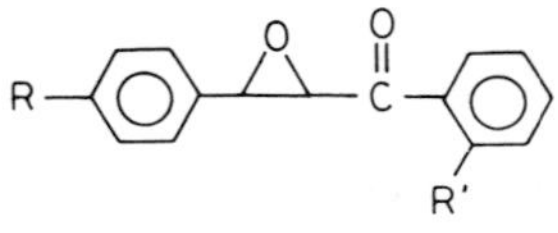

R, R′ = Me₂N, H; MeO, ɸCH₂O; Br, H; NO₂, ɸCH₂O

2-Nitrochalcone epoxide (**XII**) reacts with ethereal hydrogen chloride to yield 6-chlor-1,3-dihydroxy-2-phenylquinolin-4-(1H)-one[23] (**XIII**). The same reactants in the presence of quinol give the unchlorinated product[23] (**XIII, R = H**).

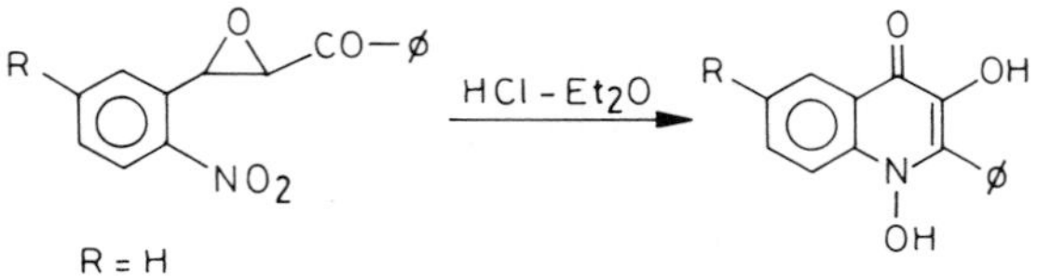

.The mechanism of this reaction can be rationalized as follows[23]:

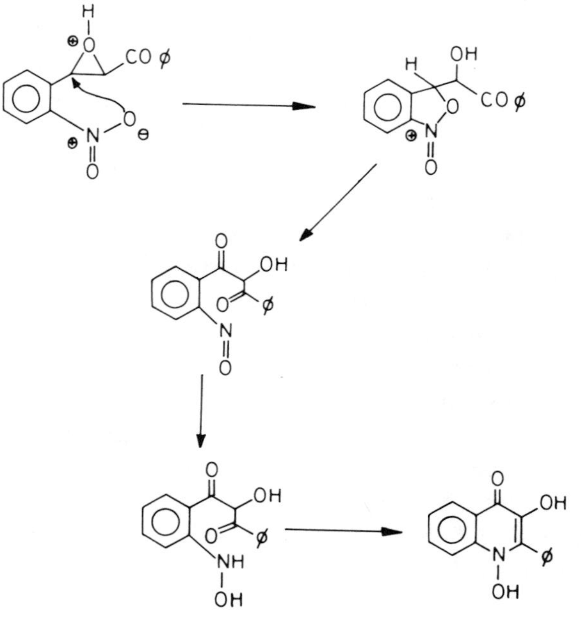

The first step in the above reaction is the protonation of the epoxide. This is followed by the opening of the oxirane ring, initiated by an attack from the oxygen (of the nitro group), resulting in the formation of nitrosophenyl ketone. The nitroso group can be reduced to the hydroxylamine group before cyclization can occur. Quinol serves this purpose, and the unchlorinated product (**XIII, R = H**) is thereby obtained. However, hydrogen chloride acting alone can cause reduction by insertion (at position 6) of the chloride ion into the original nitrophenyl nucleus (**XIII; R = Cl**).

Substituted 2′-benzyloxychalcone epoxides react with HCl saturated ether to yield either chlorohydrins or flavon-3-ol (or both), depending on the value of σ (the substitution constant)[24] of the substituent in ring A. Treatment of the epoxide of 2′-benzyloxychalcone analogue, containing condensed rings and/or a heterocyclic ring system, with HCl–Et$_2$O yield the chlorohydrins.[24]

Dilute Sulfuric Acid in Methanol[25]   *threo* and *erythro* Mixtures of **XIV** are produced by the interaction of 2′-methoxymethoxychalcone epoxide with dilute sulfuric acid and methanol:

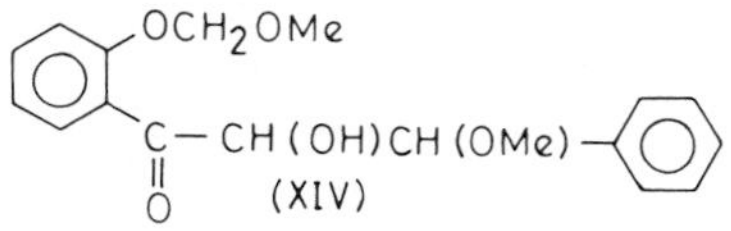

Starting with 2′-methoxymethoxy-4′,6′-dimethoxychalcone epoxide, a two-step synthesis of racemic 5,7-di-*O*-methylpinobankin has been achieved.[26]

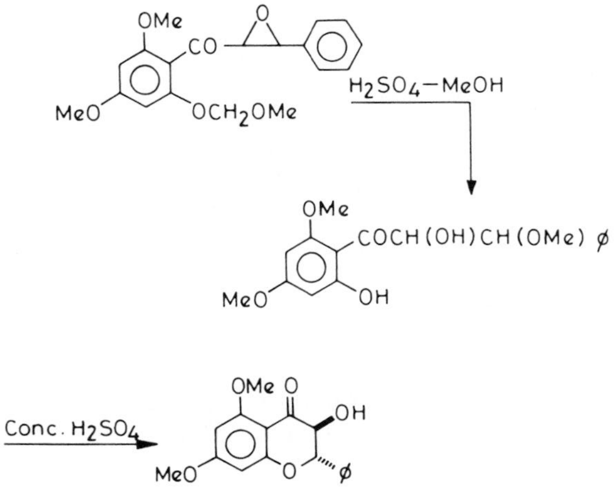

Racemic 4′,5,7-tri-*O*-methylaromadendrin and 3′,4′,5,7-tetra-*O*-methyltaxifolin have been prepared in an analogous manner.[26]

With Ethanol–HCl   2′-Benzyloxychalcone epoxide and 2′-benzyloxy-4-methoxychalcone epoxide exhibit a contrast in their behavior toward ethanolic HCl. Thus in the presence of ethanol–HCl the former epoxide undergoes cyclization to give 3-hydroxyflavanone.[27] In this example the benzyl group is split off earlier than the cleavage of the oxirane ring, and the 2′-phenolate ion formed attacks the β-carbon atom[27,28] and thus effects cyclization to 3-hydroxyflavanone. On the other hand, 2′-benzyloxy-4-methoxychalcone epoxide reacts with ethanol to give 2′-benzyloxy-4-methoxy-α-hydroxy-β-ethoxydihydrochalcone. Treatment of the latter compound with HCl results in the formation of 1-(2-benzyloxyphenyl)-2-hydroxy-3-chloro-3(*p*-methoxyphenyl)-propan-1-one.[27]

With Boron Trifluoride Etherate   5′-Substituted chalcone epoxides react with BF₃–Et₂O (including HCl) to give several products, such as, flavanonol,[19] isoflavone,[19] or chlorohydrin,[19,20] depending upon the Hammett σ value[19] of the 5′-substituent:

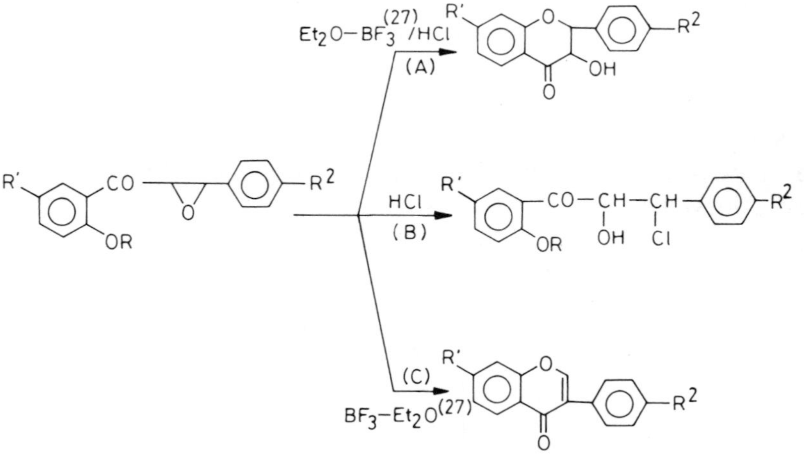

R = C₆H₅CH₂; CH₃OCH₂-
R′ = CH₃O; Cl; CH₃
R² = H; OCH₃

A: Substituent R′ σ ≤ 0
B and C: Substituent R′, σ > 0

2'-Tosyloxychalcone epoxide yields, on treatment with BF_3–Et_2O, a mixture of 3-hydroxyflavanone and flavonol[29]; on the other hand, 6'-methoxy-2'-tosyloxychalcone epoxide yields the ketoaldehyde **XV**:

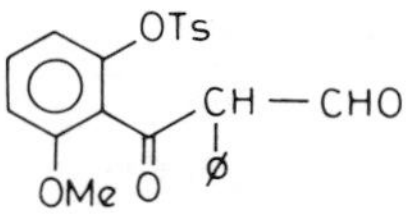

(XV)

The addition of methanol to chalcone epoxide is reported to take place according to the equation[30]:

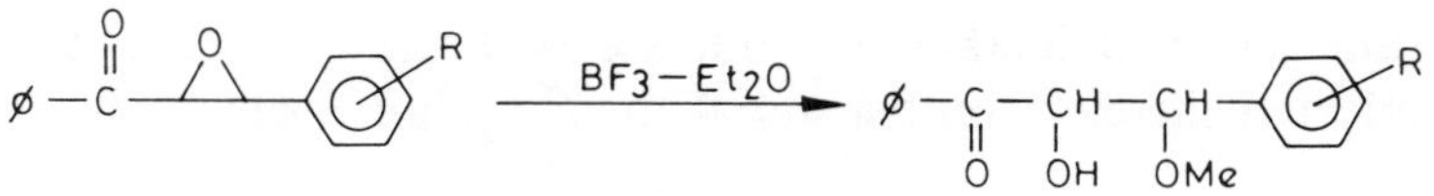

(XVI)

R = 4−OMe ; 3−NO$_2$ and 4−NO$_2$

Sometimes C_6H_5CO—$COCH_2C_6H_4$—R[30] is also formed along with the addition product.

Alkalis

2'-Tosyloxychalcone epoxide (**XVII**; R = H) with caustic alkali in methanol is reported[29] to give flavonol, while 2'-tosyloxy-6'-methoxy-chalcone epoxide (**XVII**; R = OCH_3), at room temperature or in refluxing solvent, yields 4-methoxyaurone:

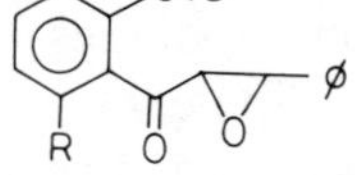

(XVII)

Stannous Chloride[31]

2-Methoxychalcone epoxide (**XVIII**) is reported to yield 1-(2-methoxybenzoyl)-2-phenylethylenechlorohydrin (**XIX**) when the benzene solution of the former is reacted with stannous chloride. **XIX** is transformed by heating (140–150°) into an α-diketone or into a diol by hydrogenation over Pd–C catalyst. Under the same conditions, **XVIII** gives a ketoalcohol.

Based on kinetic experiments the following mechanism has been reported for the ring opening of chalcone epoxides[32]:

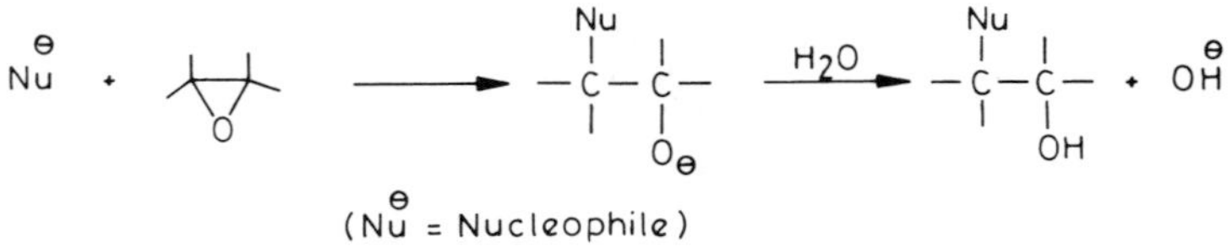

(N̄u = Nucleophile)

MISCELLANEOUS PROPERTIES

Quantum-chemical methods[28,33] indicate that the *para* substituent (B ring) effected the electron density at the β-carbon atom.

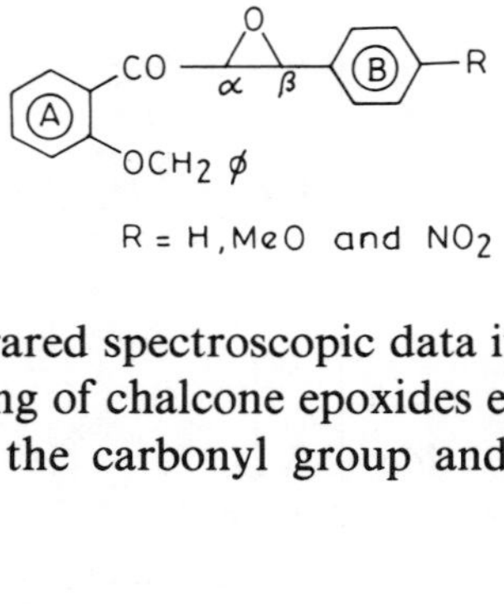

R = H, MeO and NO₂

On the basis of infrared spectroscopic data it has been demonstrated[34] that the oxirane ring of chalcone epoxides exerts an electron-withdrawing influence on the carbonyl group and thereby decreases its basicity.

CONFORMATION

The conformation of chalcone epoxides has been derived from both spectroscopic studies[34,35] and dipole moment measurements.[37] Thus a conjugated *cisoid* conformation, in the solid state as well as in CCl₄ solution, has been reported for 2'-benzyloxychalcone epoxide.[35] The *gauche* conformation[36] is adopted in solid phase by the epoxides of chalcone, 4,4'-dimethoxychalcone, and 4,4'-dichlorochalcone, which, however, change in solution (chloroform/carbon tetrachloride) to a *gauche–cis* mixture.[36] Based on dipole moment studies the following chalcone epoxides are reported to exist probably as an equilibrium mixture consisting of *gauche–syn* and *gauche–anti* conformers.[37]

$$R'-\!\!\!\bigcirc\!\!\!-CO-\!\!\!\triangle\!\!\!-\!\!\!\bigcirc\!\!\!-R$$

R = R' = H ; MeO ; and Cl

The following epoxychalcones are reported to exist in solution[38]: as equilibrium mixtures of polar *gauche* and *trans* conformers.

$$R-\!\!\!\bigcirc\!\!\!---\!\!\!\triangle\!\!\!-CO-\!\!\!\bigcirc\!\!\!-R'$$

R , R' = H,H;H,OMe ; H, Br ; Br, H ;
NO$_2$,H and H, NO$_2$

The absolute configuration of chalcone epoxide, based on chemical correlation, has appeared in the literature.[39]

REFERENCES

1 Korothkov, S. A., Orlov, V. D., and Lavrushin, V. F., *Visn. Kharkiv. Univ., Khim.*, **84**, 72 (1972); *Chem. Abstr.*, **78**, 124358c (1973).

2 Roshka, G. K., Ivanov, N. V., and Shur, A. M., *Zh. Vses. Khim. Ova.*, **19**, 703 (1974); *Chem. Abstr.*, **82**, 125192p (1975).

3 Pendse, H. K., and Limaye, S. D., *Rasayanam*, **2**, 80 (1955); *Chem. Abstr.*, **50**, 11333c (1956).

4 Litkei, G., and Bognar, R., *Kem. Kozlem.*, **34**, 249 (1970); *Chem. Abstr.*, **74**, 87706j (1971).

5 Sammour, A., Selim, M., and Hamed, A. A., *Egypt. J. Chem.*, **16**, 101 (1973); *Chem. Abstr.*, **80**, 95816j (1974).

6 Hamed, A. A., Essawy, A., and Salem, M. A., *Indian J. Chem.*, **16B**, 693 (1978).

7 Malashko, P. M., and Tishchenko, I. G., *Zh. Organ. Khim.*, **5**. 70 (1969); *Chem. Abstr.*, **70**, 87398c (1969).

8 Belyaev, V. F., and Abrazhevich, A. I., *Zh. Organ. Khim.*, **5**, 352 (1969); *Chem. Abstr.*, **70**, 114767f (1969).

9 Belyaev, V. F., and Prokopovich, V. P., *Vestsi Akad. Nauk Belarus SSR Ser. Khim. Nauk* (6), 109 (1969); *Chem. Abstr.*, **72**, 78775n (1970).

10 Litkei, G., Neubauer, A., and Bognár, R., *Magy. Kem. Foly.*, **78**, 359 (1972); *Chem. Abstr.*, **77**, 101451t (1972).

11 Neubauer, A., Litkei, G., and Bognár, R., *Tetrahedron*, **28**, 3241 (1972).

12 Dhar, D. N., Munjal, R. C., and Bose, K., *Ann. Chim. (Rome)*, **63**, 757 (1973).

13 Chopin, J., and Piccardi, G., *C. R. Acad. Sci., Paris Ser. C*, **267**, 895 (1968); *Chem. Abstr.*, **70**, 19703y (1969).

14 Piccardi, G., and Chopin, J., *Bull. Soc. Chim. France* (1), 230 (1971).

15 Cole, W., and Julian, P. L., *J. Org. Chem.*, **19**, 131 (1954).

16 Main, L., and Old, K. B., *Tetrahedron Letters*, 2809 (1977).

17 Clark-Lewis, J. W., and Tucker, P. G., *Aust. J. Chem.*, **29**, 1397 (1976).

18 Reichel, L., and Neubauer, A., Z. Chem., **8**, 423 (1968).

19 Litkei, G., and Bognár, R., *Acta Chim. (Budapest)*, **77**, 93 (1973); *Chem. Abstr.*, **79**, 78534c (1973).

20 Korotkov, S. A., Orlov, V. D., Pivnenko, N. S., and Lavrushin, V. F., *Ukr. Khim. Zh.*, **39**, 1263 (1973); *Chem. Abstr.*, **80**, 70477m (1974).

21 Sammour, A-M. A., and Elkasaby, M., *U.A.R. J. Chem.*, **13**, 151 (1970); *Chem. Abstr.*, **74**, 141176h (1971).

22 Dinya, Z., Litkei, G., Bognár, R., Matyas, G., Rochiltz, S., and Jekel, P., *Acta Chem. (Budapest)*, **77**, 323 (1973); *Chem. Abstr.*, **79**, 91383x (1973).

23 Sword, I. P., *J. Chem. Soc.*, C, 820 (1971).

24 Litkei, G., Bognár, R., and Dinya, Z., *Acta Chim. (Budapest)*, **71**, 403 (1972); *Chem. Abstr.*, **76**, 89815u (1972).

25 Pineau, J. P., and Chopin, J., *Bull. Soc. Chim. France* (10), 3678 (1971).

26 Aubry, C., and Chopin, J., *Bull. Soc. Chim. France* (12), 4503 (1971).

27 Bognár, R., and Litkei, G., *Acta Chim. (Budapest)*, **67**, 83 (1971); *Chem. Abstr.*, **74**, 99734e (1971).

28 Dinya, Z., and Litkei, G., *Magy. Kem. Foly.*, **78**, 630 (1972); *Chem. Abstr.*, **78**, 96974e (1973).

29 Gormley, T. R., and O'Sullivan, W. I., *Tetrahedron*, **29**, 369 (1973).

30 Tishchenko, I. G., Bubel, O. N., and Kononovich, V. I., *Vestn. Beloruss. Gos. Univ., Ser. 1*, **2**(2), 28 (1969); *Chem. Abstr.*, **75**, 19466b (1971).

31 Bognár, R., and Litkei, G., *Acta Univ. Debrecen. Ludovico Kossuth Nom.*, **8**, 25 (1962); *Chem. Abstr.*, **59**, 13924f (1963).

32 Chanda, B. K., Kar, J. N., Behera, G. B., and Rout, M. K., *J. Indian Chem. Soc.*, **48**, 867 (1971).

33 Dinya, Z., and Litkei, G., *Acta Chim. (Budapest)*, **77**, 211 (1973); *Chem. Abstr.*, **78**, 65571n (1973).

34 Orlov, V. D., Korotkov, S. A., Sukach, Y. A., and Lavrushin, V. F., *Zh. Obshch. Khim.*, **43**, 1353 (1973).

35 Dinya, Z., Litkei, G., Bognár, R., and Matayas, G., *Magy. Kem. Foly.*, **78**, 504 (1972); *Chem. Abstr.*, **78**, 42310s (1973).

36 Orlov, V. D., Korotkov, S. A., Sukach, Y. A., and Lavrushin, V. F., *Khim. Geterotsikl. Soedin.* (3), 306 (1975); *Chem. Abstr.*, **82**, 169880r (1975).

37 Vereshchagin, A. N., Vul'fson, S. G., Donskova, A. I., and Savin, V. I., *Dokl. Akad. Nauk SSSR*, **215**, 339 (1974); *Chem. Abstr.*, **81**, 3258y (1974).

38 Arbuzov, B. A., Donskova, A. I., Vul'fson, S. G., and Vereshchagin, A. N., *Izv. Akad. Nauk SSSR, Ser. Khim.* (7), 1498 (1975); *Chem. Abstr.*, **83**, 178111p (1975).

39 Marsman, B., and Wynberg, H., *J. Org. Chem.*, **44**, 2312 (1979).

Chapter Twenty-Eight
Chalcone α,β-Dibromides

REACTIONS

The reactions of substituted chalcone dibromides have been studied in considerable detail. Thus under varied experimental conditions the following types of compounds have been obtained from chalcone dibromides: flavones,[1-23] flavonol,[24,28] flavanones,[29,30] aurones,[2,7,8,10,11,22,24-26]

benzoylcoumarones,[12] α-bromochalcones,[22,31] tetralones,[33] and aziridines.[5,23]

Flavones[1–23]

Flavones have been secured either by the pyrolysis[3,9] of o-hydroxy(or
o-acetoxy-) chalcone α,β-dibromides or by their reaction with alcoholic alkali,[2,5–13] sodium methoxide,[4] pyridine,[1,15–21] or potassium cyanide.[3,22]
The latter reaction has been utilized for the synthesis of naturally occurring flavones, containing a phloroglucinol nucleus, viz., chrysin,
apigenin, and luteolin.[3]

In the conversion of o-substituted chalcone dibromide into the corresponding flavone, under the influence of a base, the intermediate formation of a β-methoxy compound has been postulated. The following
is a typical example[4]:

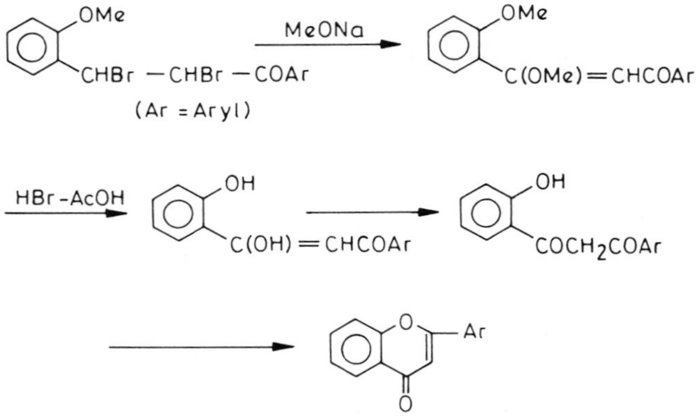

The formation of flavone from 2'-hydroxy-3-nitrochalcone dibromide
has been rationalized in the following way[8]:

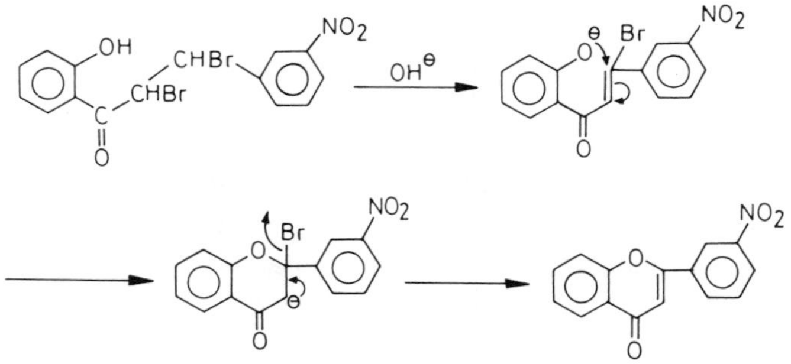

On refluxing 2'-hydroxy-3'-methyl-5'-nitro-4-methoxychalcone dibromide with pyridine the corresponding flavone is obtained.[1] In this case pyridine behaves like an alkali and brings about dehydrobromination and cyclization. Nuclear substitution in this case is not feasible, since the probable position of entry of the bromine is already occupied. Utilizing this reaction some substituted flavones have been synthesized. 8-Nitro-7-hydroxyflavone has been prepared in the aforesaid manner.[15] The isomeric compound 5-hydroxy-6-nitroflavone, however, can be prepared[15] as follows:

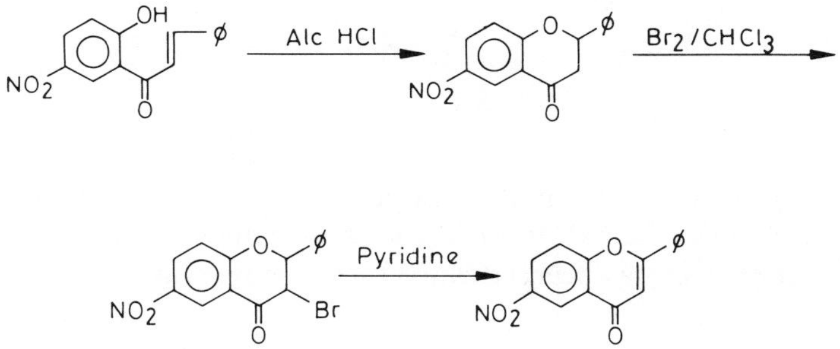

In the synthesis of flavones from chalcone dibromides and ethanolic alkali it has been demonstrated that steric effects do play an important role.[13] Thus flavone derived from chalcone (**Ia**) is obtained in 89% yield, while the flavone from chalcone (**Ib**) is secured in a relatively low yield (46%). In this reaction aurone is also formed, but in smaller amounts.[13]

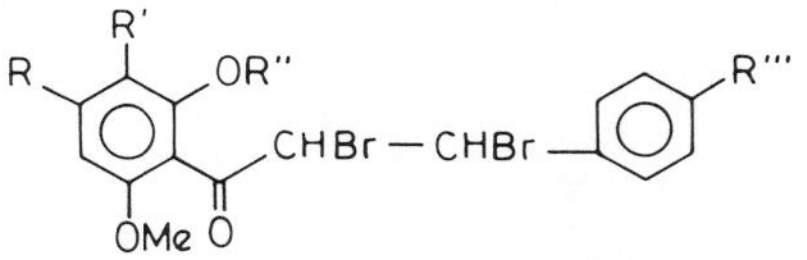

(Iа) R = R' = R'' = H; R''' = Ac
(Ib) R = R'' = H; R' = Br; R''' = Ac

A mixture of 2'-hydroxy-5'-methyl-4-methoxychalcone dibromide (**II**) and 2'-hydroxy-5'-methylchalcone (**III**) is reported to react in the presence of pyridine, leading to the formation of cross-brominated products, **IV** and **V**[23]:

The formation of these products arises due to debromination and nuclear monosubstitution by an intermolecular mechanism, followed by oxidative cyclization.

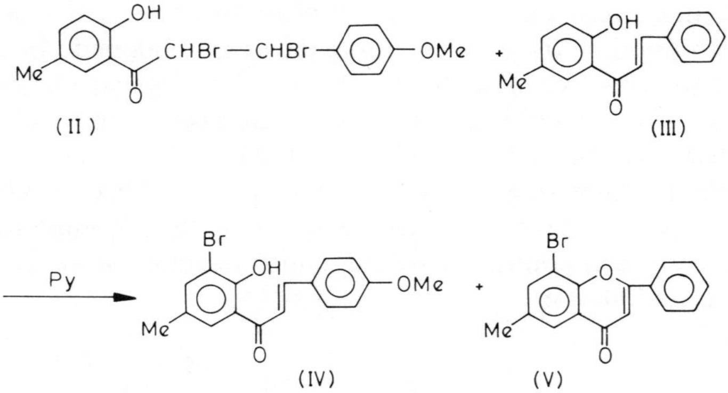

Aurones[2,7,8,10,11,22,24,25,26]

2'-Hydroxy-4-nitrochalcone dibromide undergoes ring cyclization under the influence of alkali to yield the corresponding aurone. The following mechanism has been advanced for this reaction[8]:

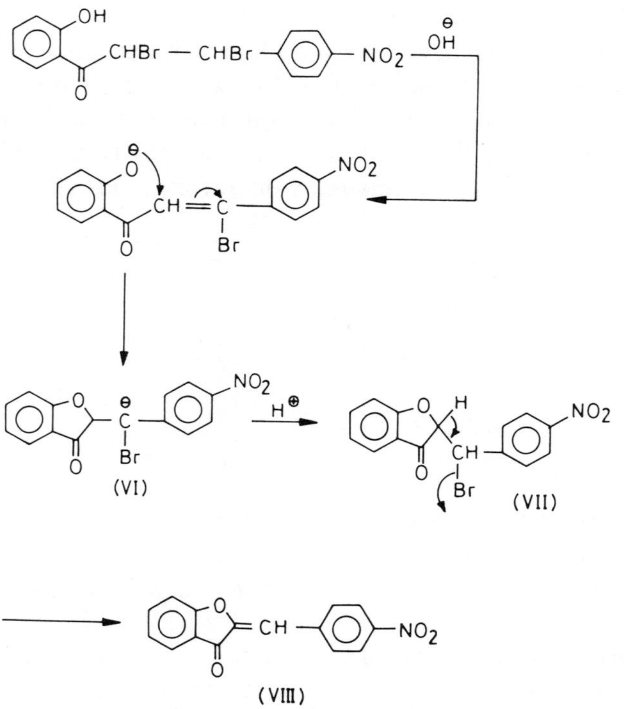

The first step in the reaction is the formation of β-bromochalcone, which then furnishes the anion (**VI**). The anion gets protonated, and the aurone hydrobromide (**VII**) loses a molecule of HBr to form the aurone (**VIII**).

Flavones are generally obtained from 2'-acetoxychalcone dibromides and dichlorides, having a substituent in the 6'-position.[7] Aurones, however, are formed when the dihalides carry substituents in both the 3'- and 6'-positions.[7] Apparently a steric effect plays an important part in the formation of aurone.[7]

Aurones are produced along with flavones in the reaction of chalcone dibromides with alcoholic potassium cyanide[22] or dilute alcoholic alkali.[2,10,11] The ratio of aurone to flavone in the latter case is reported to increase with increasing concentration of the alkali.[10,11] The following mechanism has been suggested for the simultaneous formation of aurone and flavone from the chalcone dibromide.[10,11]

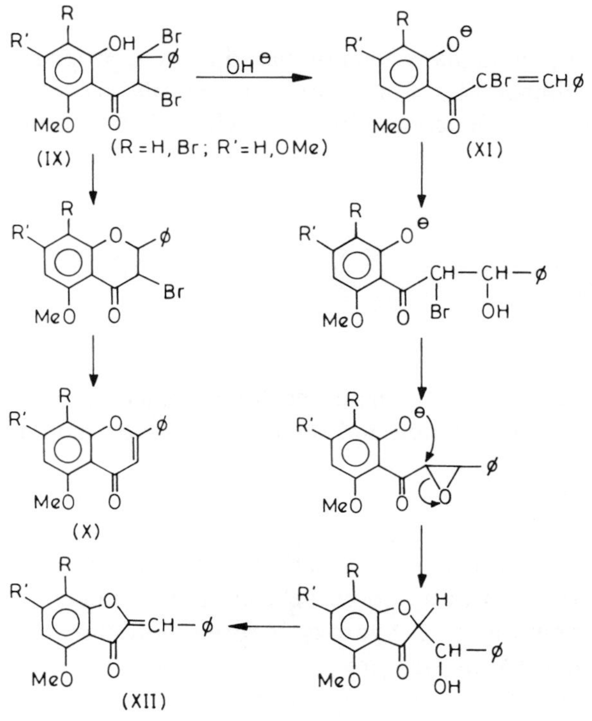

The intermediate α-bromochalcone (**XI**) exhibits geometrical isomerism. Its *trans* isomer cyclizes more readily to the corresponding aurone[11] in the presence of a base.

Benzoylcoumarone

Some of the o-acetoxychalcone dibromides on treatment with a base yield the corresponding benzoylcoumarones. The following example is illustrative[12]:

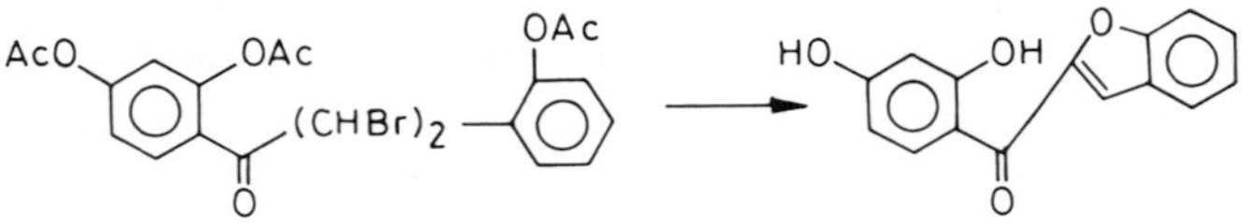

2',4'- Dihydroxy - 2 - benzoyl - Coumarone

Flavonol and Dihydroflavonols[27,28]

The title compounds have been prepared from appropriately substituted chalcone dibromides. Flavonol, [27] for example, is obtained by the reactionof o-acetoxychalcone dibromide with alkali in aqueous–acetone medium:

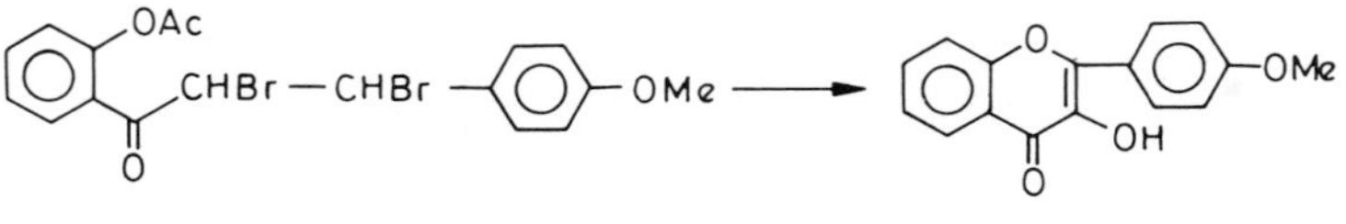

It is postulated that the aqueous–acetone replace the α-bromine atom with the hydroxyl function and removes two hydrogen atoms.[28]

Flavanones

Treatment of 2'-acetoxy-4,6'-dimethoxychalcone dibromide with acetic acid results in deacetylation, debromination, nuclear bromination, and the formation of 2'-hydroxy-3'-bromo-4,6'-dimethoxychalcone[29] (**XIV**) and not **XV** as was previously reported.[30] Then **XIV** smoothly cyclizes to the corresponding flavanone,[29] **XVI**.

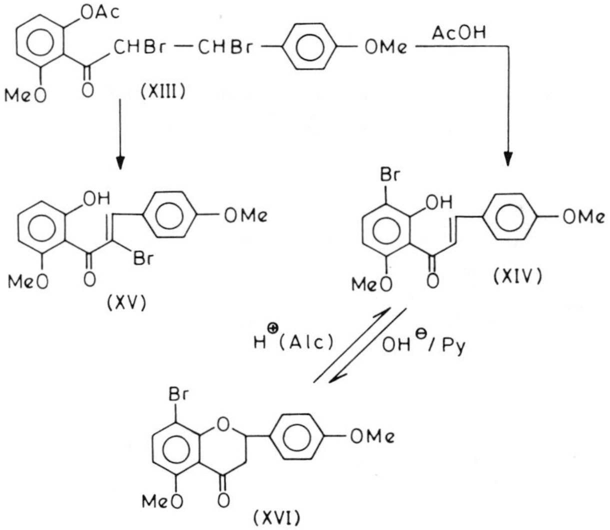

Under appropriate experimental conditions chalcone dibromides can be transformed into a variety of products, including α-bromochalcones,[22,31] aziridines,[5] and aminoflavanones.[5] α-Bromochalcones[31] are obtainable by dehydrobromination of chalcone dibromides brought about by alcoholic potassium hydroxide. α-Chlorochalcones, however, can be obtained by dehydrochlorination with potassium acetate in absolute ethanol.[32]

The preparation of aminoflavanones (**XVIII**) via the aziridine intermediate (**XVII**) has been accomplished by using 2′-substituted chalcone dibromide as the starting material.[5] Thus

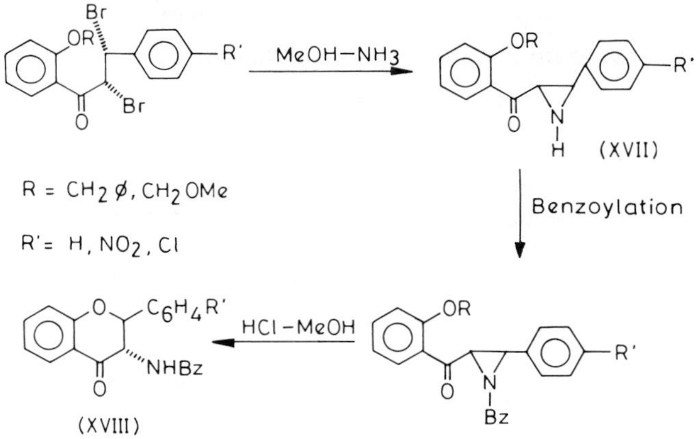

$R = CH_2\phi, CH_2OMe$

$R' = H, NO_2, Cl$

Tetralones

Chalcone dibromides have been utilzed as starting materials for the prepartion of 2-aryl-1-tetralone[33] (**XIX**) by the following sequence of reactions:

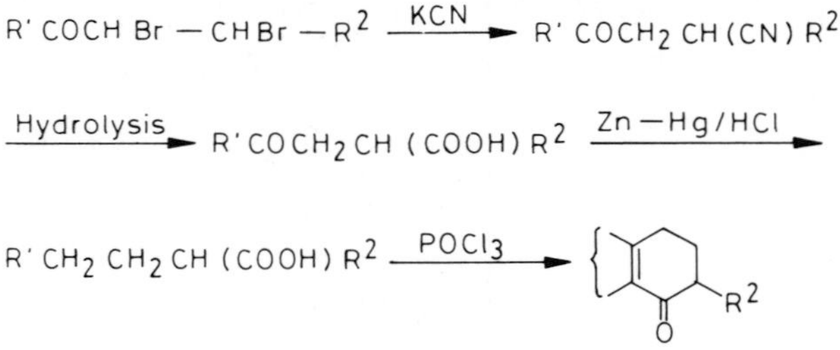

$$R'COCHBr-CHBr-R^2 \xrightarrow{KCN} R'COCH_2CH(CN)R^2$$

$$\xrightarrow{Hydrolysis} R'COCH_2CH(COOH)R^2 \xrightarrow{Zn-Hg/HCl}$$

$$R'CH_2CH_2CH(COOH)R^2 \xrightarrow{POCl_3}$$

(R' and R^2 are aryl groups)   (XIX)

Benzalacetone

According to a patent the synthesis of benzalacetone[34] has been effected by the reaction of chalcone dibromide with alkali mercaptide.

$$\phi(CHBr)_2-CO-\phi \xrightarrow{Na/MeOH,\ PrSH} \phi CH=CH\ COMe$$

Aziridines[35,36]

Aziridines have been formed by the reaction between $R'NH_2$ ($R' = H$, Me, or Bz) and 4-nitrochalcone dibromide.[36] The following mechanism is postulated to be involved in this transformation[36]:

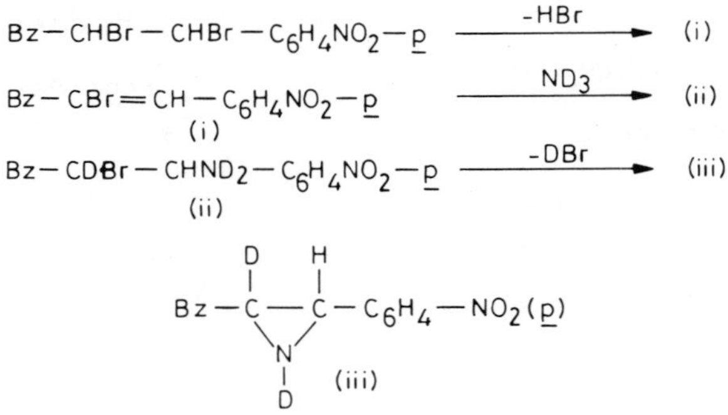

$$Bz-CHBr-CHBr-C_6H_4NO_2-\underline{p} \xrightarrow{-HBr} \quad (i)$$

$$Bz-CBr=CH-C_6H_4NO_2-\underline{p} \xrightarrow{ND_3} \quad (ii)$$
$$(i)$$

$$Bz-CDBr-CHND_2-C_6H_4NO_2-\underline{p} \xrightarrow{-DBr} \quad (iii)$$
$$(ii)$$

$$Bz-\underset{\underset{N}{|}}{\overset{\overset{D}{|}}{C}}-\underset{\underset{|}{H}}{\overset{\overset{H}{|}}{C}}-C_6H_4-NO_2(\underline{p})$$
$$\underset{D}{|}\quad (iii)$$

MISCELLANEOUS REACTIONS

Halogens, Alcohols, and Potassium Cyanide

The additional reactions of chalcone dihalides[37] are summarized below:

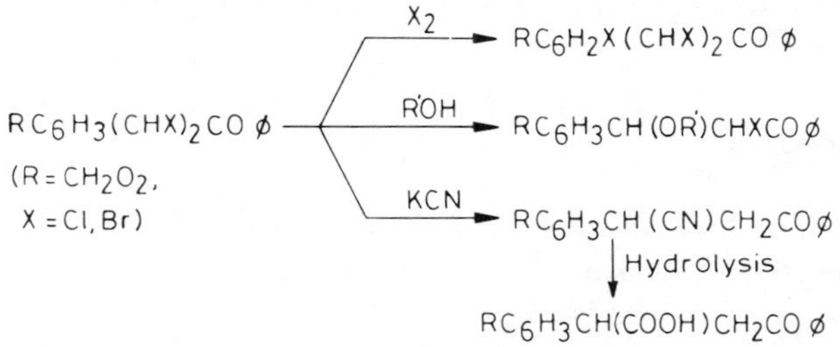

$$RC_6H_3(CHX)_2CO\,\phi \quad (R = CH_2O_2,\; X = Cl, Br)$$

$$\xrightarrow{X_2} RC_6H_2X(CHX)_2CO\,\phi$$

$$\xrightarrow{R'OH} RC_6H_3CH(OR')CHXCO\,\phi$$

$$\xrightarrow{KCN} RC_6H_3CH(CN)CH_2CO\,\phi$$

$$\xrightarrow{Hydrolysis} RC_6H_3CH(COOH)CH_2CO\,\phi$$

Hydrazine Hydrate

Treatment of dibromochalcone with hydrazine hydrate yields 3,5-diphenylpyrazole.[38] The reactions involved in these conversions are summarized below[38]:

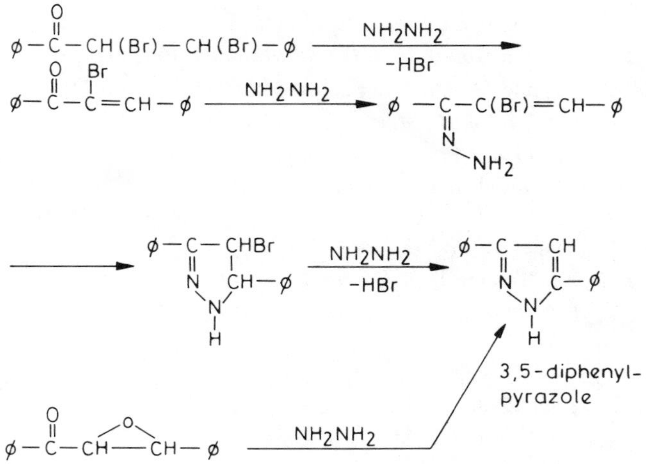

Sodium Azide[39]

2'-Hydroxy- (and 2'-benzyloxy-) chalcone dibromides are reported to react with sodium azide, leading to the formation of α-azidochalcones.[39]

Cleavage of α-Bromochalcone with HBr and Oxygen

α-Bromochalcone (obtained by the dehydrobromination of α,β-dibromochalcones) is reported to react with HBr and oxygen in petroleum ether, yielding benzoic acid.[40] 4'-Methyl-α-bromochalcone yields *p*-toluic acid on similar treatment. These two examples illustrate a novel cleavage reaction brought about through the oxygen effect.

REFERENCES

1   Wadodkar, P. N., *Indian J. Chem.*, **1**, 163 (1963).

2   Nadkarni, S. M., Warriar, A. M., and Wheeler, T. S., *J. Chem. Soc.*, 1978 (1937).

3   Hutchinson, W. A., and Wheeler, T. S., *J. Chem. Soc.*, 91 (1939).

4   Bhagwat, N. A., and Wheeler, T. S., *J. Chem. Soc.*, 94 (1939).

5   Litkei, G., Bognár, R., and Ando, J., *Acta Chim.* (*Budapest*), **76**, 95 (1973); *Chem. Abstr.*, **79**, 18533s (1973).

6   Litkei, G., and Bognár, R., *Acta Phys. Chem. Debrecina*, **17**, 239 (1971); *Chem. Abstr.*, **78**, 42970g (1973).

7   Donnelly, J. A., Doran, H. J., and Murphy, J. J., *Tetrahedron*, **29**, 1037 (1973).

8   Donnelly, D. J., Donnelly, J. A., and Philbin, E. M., *Tetrahedron*, **28**, 1867 (1972).

9   Donnelly, D. J., Donnelly, J. A., and Philbin, E. M., *Tetrahedron*, **28**, 53 (1972).

10 Donnelly, J. A., and Doran, H. J., *Tetrahedron*, **31**, 1565 (1975).

11 Donnelly, J. A., and Doran, H. J., *Tetrahedron*, **31**, 1791 (1975).

12 Tambor, J., and Gubler, H., *Helv. Chim. Acta*, **2**, 101 (1919).

13 Donnelly, D. J., Donnelly, J. A., Murphy, J. J., Philbin, E. M., and Wheeler, T. S., *Chem. Commun.* (12), 351 (1966).

14 Chhaya, G. S., Trivedi, P. L., and Jadhav, G. V., *J. Univ. Bombay*, **27A**, Part 3, 26 (1958); *Chem. Abstr.*, **54**, 8807e (1960).

15 Seshadri, S., and Trivedi, P. L., *J. Org. Chem.*, **23**, 1735 (1958).

16 Marathey, M. G., *J. Univ. Poona, Sci. Technol.*, **18**, 53 (1960); *Chem. Abstr.*, **55**, 1598e (1961).

17 Naik, V. G., and Marathey, M. G., *J. Univ. Poona, Sci. Technol.*, **18**, 55 (1960); *Chem. Abstr.*, **55**, 1598e (1961).

18 Naik, V. G., and Marathey, M. G., *J. Univ. Poona, Sci. Technol.*, **18**, 61 (1960); *Chem. Abstr.*, **55**, 1599a (1961).

19 Gore, K. G., Naik, V. G., and Marathey, M. G., *J. Univ. Poona, Sci. Technol.*, **18**, 65 (1960); *Chem. Abstr.*, **55**, 1599e (1961).

20 Gore, K. G., Naik, V. G., and Marathey, M. G., *J. Univ. Poona, Sci. Technol.*, **18**, 69 (1960); *Chem. Abstr.*, **55**, 1599h (1961).

21 Gore, K. G., Somaware, H. R., Thakar, G. P., and Marathey, M. G., *J. Univ. Poona, Sci. Technol.*, **18**, 73 (1960); *Chem. Abstr.*, **55**, 1600c (1961).

22 Vandrewalla, H. P., and Jadhav, G. V., *Proc. Indian Acad. Sci.*, **28A**, 125 (1949).

23 Ghiya, B. J., and Marathey, M. G., *Indian J. Chem.*, **6**, 766 (1968).

24 Khanolkar, A. P., and Wheeler, T. S., *J. Chem. Soc.*, 2118 (1938).

25 Joshi, K. C., and Jauhar, A. K., *Indian J. Chem.*, **1**, 477 (1963).

26 Donnelly, D. J., Donnelly, J. A., and Keegan, J. R., *Tetrahedron*, **33**, 3289 (1977).

27 Marathey, M. G., *J. Univ. Poona, Sci. Technol.*, No. 4, 73 (1953); *Chem. Abstr.*, **49**, 10944i (1955).

28 Marathey, M. G., *Sci. Culture*, **16**, 527 (1951).

29 Donnelly, J. A., *Tetrahedron*, **29**, 2585 (1973).

30 Pendse, H. K., *Rasāyanam*, **2**, 131 (1956); *Chem. Abstr.*, **51**, 5063f (1957).

31 Sharma, T. C., Patel, H., and Bokadia, M. M., *Indian J. Chem.*, **11**, 703 (1973).

32 Brosche, K., Weber, F. G., Westphal, G., and Reimann, E., *Z. Chem.*, **19**, 96 (1979).

33 Hidayetulla, M. S., Shah, R. C., and Wheeler, T. S., *J. Chem. Soc.*, 111 (1941).

34 Thompson, R. B., U.S. Patent 2,553,797 (1951); *Chem. Abstr.*, **46**, 7579e (1952).

35 Tarburton, P., Wolpa, L. J., Loerch, R. K., Folsom, T. L., and Cromwell, N. H., *J. Heterocycl. Chem.*, **14**, 1203 (1977).

36 Weber, F. G., and Bandlow, C., *Z. Chem.*, **13**, 467 (1973).

37 Dodwadmath, R. P., and Wheeler, T. S., *Proc. Indian Acad. Sci.*, **2A**, 438 (1935); *Chem. Abstr.*, **30**, 1770⁹ (1936).

38 Sharma, T. C., Saxena, M. K., and Bokadia, M. M. *Indian J. Chem.*, **9**, 794 (1971).

39 Patonay, T., Rakesi, M., Litkei, G., Mester, T., and Bognár, R., *Flavonoids Bioflavonoids, Proc. Hung. Bioflavonoid Symp., 5th*, 227 (1977); *Chem. Abstr.*, **89**, 108954r (1978).

40 Kasiwagi, V. H., *Bull. Chem. Soc. Japan*, **26**, 355 (1953).

Chapter Twenty-Nine

Uses

Many patents have appeared in the literature describing the usefulness of chalcones and their derivatives. These find application as artificial sweeteners,[1-20] stabilizer[21-40] against heat, visible light, ultraviolet light, aging, color photography,[55] scintillators,[65] polymerization catalysts,[68,69] fluorescent whitening agents,[70] and organic brightening additives.[71,72]

SWEETENERS[1-20]

The chalcone derivatives dihydrochalcones[1] and their corresponding glycosides have been employed as food-sweetening agents. Dihydrochalcone xylosides[2] and galactoside,[2,3,10] for example, are claimed to be 1.5–2 times sweeter than saccharin. Incorporation of cyclodextrin with the sweetener neohesperidindihydrochalcone,[6,7] is reported to stabilize its aqueous solution.[11] The preparation of a mixture of glucosides[5,8,9] (mono-, di-, tri-, tetra-, and penta-) of hesperidindihydrochalcone[4-6] has been described. 3,2',4',6'-Tetrahydroxy-4-propoxy-dihydrochalcone-4β'-neohesperdoside[12] has been used as a synthetic sweetener and is 2200 times sweeter than glucose.

STABILIZERS

2',4-Dihydroxy-3,5-di-*tert*-butylchalcone has been employed as an oxidation inhibitor[21] and stabilizer[22] to polyproplene polymer.

Hydroxychalcones form the starting material for the preparation of hydroxyflavones, which serve as antioxidants for lipoid material.[23]

Chalcone is a natural constituent of beer and plays, in combination with other polyphenols, an important role in its stability.[24]

Chalcone forms the constituent of corrosion-inhibiting lubricants suited for internal combustion engines containing silver and similar metal components.[25] According to one patent the efficiency of lubricant additive is retained by incorporation of chalcone, otherwise it is diminished by reaction of the additive with olefinic components of base oils or grease.[26]

The incorporation of 2',4,4'-trimethoxychalcone into pulp sheets (on which pesticide was absorbed) helped in retarding the air degradation.[27]

It has been reported that the addition (up to 5%) of chalcone to poly(dimethoxysiloxane), the silicone dielectric fluid, for impregnation of capacitors, etc., improves its dielectric life.[28] Chalcone[29] and β-(benzoyloxy)-2'-hydroxychalcone[30] have proved to be good light absorbers and heat stabilizers for polymeric materials, for example, polymethyl methacrylate film and PVC resin sheets. Organic esters having a chalcone-type skeleton, $RC_6H_3(OH)COCH=C(OOCC_6H_5)C_6H_5$ [where R = Br or H], have been employed for the aforesaid purpose.[31]

Incorporation of 0.02–5% chalcone is claimed to prevent discoloration of microcrystalline petroleum waxes[32] and polymers [33] (halogen-containing polyvinyl compounds) exposed to sunlight. Likewise polyolefins are stabilized[34] against light, heat, and aging by adding small amounts of chalcone.

Chalcone has been described to possess UV absorption property and hence finds application as UV absorption filters.[35,36]

Chalcone serves as a suitable ultraviolet absorption additive in adhesives, lacquers, and plastics.[37]

A nonirritating preparation containing 1–5% chalcone for the protection[38,39] of skin from sunlight has been patented.

It is reported that the properties of cellulosic material are improved with ionization radiation in the presence of a graftable organic finishing agent, consisting of chalcone and a sensitizer.[40]

PHOTOSENSITIVE MATERIALS

A number of chalcone derivatives[22,41–47] form the principal ingredients in the preparation of photosensitive polymeric material, some of which possess good film-forming properties.[42]

Photocrosslinkable copolymers[48] are used for the preparation of sharp relief images and printing blocks. Photoresist compositions have been developed for the preparation of printed multilayer circuits. The printed circuit image is claimed to have a high edge sharpness.[49]

Light-sensitive preparations useful for printing plates of improved sensitivity and resolution are described, which, for example, involve the reaction of 4'-(2-hydroxyethoxy)chalcone with styrene–maleic anhydride copolymer under appropriate conditions.[50–53]

Another light-sensitive resin has been reported,[54] which possesses good adhesion, toughness, alkali resistance, and stability toward oxidation. The resin has been obtained by heating 3,4'-dihydroxychalcone with an epoxyresin in a suitable solvent, in the presence of alkali.[54]

Chalcone, 4-isocyanatochalcone, and furan analogue of chalcone have proved useful in the preparation of light-sensitive film used in color photography.[55] Chalcone is also used in the preparation of photothermographic emulsion.[56]

4,4-Diphenylaminochalcone[57] and 4,4'-bis(diphenylamino)chalcone[58,59] have been used as constituents of a photoconducting composition for use in electrophotographic products.

Photographic silver halide emulsion spectrally sensitized with carbocyanine dyes, can be supersensitized[60] with some chalcones, for example, 2'- (and 4'-) chlorochalcones.

POLYMERS

Dihydroxychalcone has been used[61,62] for the preparation of uncured epoxyresins. 4,4'-Dihydroxychalcone forms the component of a duroplastic mixture,[63] which possessess good mechanical properties and a high thermal stability.

Methacryloyloxychalcones are used as crosslinking agents[64] in the preparation of butyl acrylate–styrene copolymers, which are claimed[64] to be useful as lacquers.

SCINTILLATORS

2,4,6-Trisubstituted pyridines, derived from chalcones and formamide (Leuckart reaction), exhibit extremely intense fluorescence and could be used as scintillators.[65]

ANALYTICAL REAGENTS

Chalcones react with a number of metal ions and are reported to be more reactive than the aldehyde or ketone from which they have been prepared.[66] This reaction has been exploited.[67] for the detection of Fe^{3+} (limit of identification: $0.33\gamma/0.05$ ml) by 2',4'-dihydroxychalcone, provided the concentration of interfering ions is kept at a minimum.

MISCELLANEOUS APPLICATIONS

Polymerization Catalysts, Fluorescent Whitening Agents, and Organic Brightening Additives

Chalcone is reported to form a component of a polymerization catalyst designed for obtaining highly crystalline polyolefinic polymers, for example, polyacrylates, in high yields.[68,69]

Chalcone sulfonic acids serve as intermediates in the preparation of fluorescent whitening agents.[70] Chalcones have also been employed as organic brightening additives.[71,72]

REFERENCES

1   Linke, H. A. B., and Eveleigh, D. E., *Z. Naturforsch.*, **30B**, 740 (1975).

2   Horowitz, R. M., and Gentili, B., U.S. Patent, 3,890,298 (1975); *Chem. Abstr.*, **83**, 147706g (1975).

3   Horowitz, R. M., and Gentili, B., U.S. Patent, 3,890,296 (1975); *Chem. Abstr.*, **83**, 147703d (1975).

4   Rizzi, G. P., and Neely, J. S., German Patent, 2,148,332 (1972); *Chem. Abstr.*, **77**, 86777h (1972).

5   Fukomoto, J., and Okada, S., German Patent, 2,204,716 (1972); *Chem. Abstr.*, **77**, 152515z (1972).

6   Horowitz, R. M., and Gentili, B., *Sweetness Sweeteners, Proc. Ind. Univ. Co-op. Symp.*, 69 (1971); *Chem. Abstr.*, **78**, 56483n (1973).

7   Robertson, G. H., Clark, J. P., and Lundin, R., *Ind. Eng. Chem., Prod. Res. Develop.*, **13**, 125 (1974); *Chem. Abstr.*, **81**, 37745v (1974).

8   Horowitz, R. M., and Gentili, B., U.S. Patent, 3,429,873 (1969); *Chem. Abstr.*, **70**, 87315y (1969).

9   Laboratoires Coupin (S.A.), French Patent, 1,221,869 (1960); *Chem. Abstr.*, **56**, P 10046i (1962).

10 Horowitz, R. M., and Gentili, B., U.S. Patent 3,876,777 (1975); *Chem. Abstr.*, **83**, 28539v (1975).

11 Hashimoto, S., and Katayama, A., Japanese Patent 75 35,349 (1975); *Chem. Abstr.*, **83**, 7607a (1975).

12 International Minerals and Chemical Corpn., British Patent 1,189,573 (1970); *Chem. Abstr.*, **73**, 45798q (1970).

13 Chen, C. I., Lin, Y-Y, and Lin, Y-C., *T'ai-wan K'o Hsueh*, **28**, 40 (1974); *Chem. Abstr.*, **83**, 42971m (1975).

14 Lindner, K., Czelenyi, C., Kubat, K., Bolla, F., and Szejtli, J., *Acta Aliment. Acad. Sci. Hung.*, **6**, 311 (1977); *Chem. Abstr.*, **88**, 103430m (1978).

15 Antus, S., Farkas, L., Gottsegen, A., Nogradi, M., Strelisky, J., and Pfliegel, T., *Acta Chim. Acad. Sci. Hung.*, **98**, 231 (1978); *Chem. Abstr.*, **90**, 121124k (1979).

16 Antus, S., Farkas, L., Gottsegen, A., Nogradi, M., and Pfliegel, T., *Acta Chim. Acad. Sci. Hung.*, **98**, 225 (1978); Chem. Abstr., **90**, 86935b (1979).

17 Linke, H. A. B., and Eveleigh, D. E., U.S. Patent 4,087,558 (1978); *Chem. Abstr.*, **89**, 90260z (1978).

18 Kamiya, S., Ezaki, S., Konishi, F., and Watanable, T., Japan Kokai, 78 37,646 (1978); *Chem. Abstr.*, **89**, 110284r (1978).

19 Kamiya, S., Konishi, F., and Ezaki, S., *Agric. Biol. Chem.*, **42**, 941 (1978).

20 Yamato, M., Hashigaki, K., Mito, K., and Koyama, T., *Chem. Pharm. Bull.*, **26**, 2321 (1978); *Chem. Abstr.*, **90**, 22748k (1979).

21 Adams, J. H., British Patent, 1,250,388 (1971); *Chem. Abstr.*, **76**, 46956p (1972).

22 Akhmedzade, D. A., Yasnopol'skii, V. D., and Golovanova, Y. I., *Azerb. Khim. Zh.* (2), 117 (1968); *Chem. Abstr.*, **69**, 107333n (1968).

23 Simpson, T. H. and Uri, N., British Patent, 875,164 (1956); *Chem. Abstr.*, **56**, P 3582b (1962).

24 Karel, V., *Brauwissenschaft*, **14**, 411 (1961); *Chem. Abstr.*, **56**, 5220a (1962).

25 Fields, E. K., U.S. Patent 2,799,652 (1957); *Chem. Abstr.*, **51**, 15113c (1957).

26 Bayer, F., German Patent 919,128 (1954); *Chem. Abstr.*, **52**, 13244i (1958).

27 Aries, R., French Patent 2,129,811 (1972); Addn. to French Patent 2,096,713; *Chem. Abstr.*, **78**, 120265x (1973).

28 General Electric Company, German Patent, 1,106,821 (1961); *Chem. Abstr.*, **57**, 5460i (1962).

29 Gunder, O. A., Vlasov, V. G., Koval, L. N., and Krasovitskii, B. M. *(USSR), Plast. Massy* (6), 3 (1968); *Chem. Abstr.*, **69**, 36699y (1968).

30 Buckman, S. J., Flanagan, K. J., Pera, J. D., and Wienert, L. A., U.S. Patent 3,629,322 (1971); *Chem. Abstr.*, **76**, 114277v (1972).

31 Buckman, S. J., Flanagan, K. J., Pera, J. D., and Wienert, L. A., S. African Patent 6902,541 (1969); *Chem. Abstr.*, **72**, 122440a (1970).

32 Rumberger, G. G., U.S. Patent 2,755,193 (1956); *Chem. Abstr.*, **50**, P 16101e (1956).

33 Chemische werke Huls G.m.b.H., British Patent, 710,964 (1954); *Chem. Abstr.*, **49**, 668e (1955).

34 Bonvicini, A., "Montecantini," Societa Generale per l industria' Mineraria e chimica. Italian Patent, 662,500 (1964); *Chem. Abstr.*, **63**, 1950c (1965).

35 Ikeda, H., and Toba, K., *Rept. Govt. Chem. Ind. Research Inst., Tokyo*, **51**, 65 (1956); *Chem. Abstr.*, **50**, 11110i (1956).

36 Ikeda, H., and Toba, K., *J. Soc. Sci. Phot. Japan*, **18**, 110 (1956); *Chem. Abstr.*, **51**, 2395e (1957).

37 Stecher, H., *Adhaesion*, **2**, No. 6, 243 (1958); *Chem. Abstr.*, **53**, 14579c (1959).

38 Giese, A. C., Christensen, E., and Jeppson, J., *J. Am. Pharm. Assoc.*, **39**, 30 (1950); *Chem. Abstr.*, **44**, 3209b (1950).

39 Fischer, F., Schuchhardt, W., and Walter, H., East German Patent 69,672 (1969); *Chem. Abstr.*, **73**, 7093y (1970).

40 Munzel, F., U.S. Patent, 3,254,939 (1966); *Chem. Abstr.*, **65**, 7358b (1966).

41 Hatanaka, H., Sugiyama, K., Nakaya, T., and Imoto, M., *Mikromol. Chem.*, **176**, 3231 (1975); *Chem. Abstr.*, **84**, 31790w (1976).

42 Ichibashi, T., and Kawai, W. Japanese Patent 7414,352 (1974); *Chem. Abstr.*, **82**, 17466x (1975).

43 Ichibashi, T., and Kawai, W., Japanese Patent 7501,191 (1975); *Chem. Abstr.*, **83**, 35721b (1975).

44 Kato, M., Hasegawa, M., and Ichijo, T., U.S. Patent, 3,873,500 (1975); *Chem. Abstr.*, **83**, 29065z (1975).

45 Nakane, H., Ichii, K., Ayabe, T., and Asaumi, S., Japanese Patent 72 39,184 (1972); *Chem. Abstr.*, **78**, 112129r (1973).

46 Panda, S. P., *J. Appl. Polymer. Sci.*, **18**, 2317 (1974).

47 Panda, S. P., *J. Armament Stud.*, **11**, 30 (1975).

48 Ehrig, B., Mueller, E., Mott, L., and Wolf, E., German Patent 2,116,010 (1972); *Chem. Abstr.*, **78**, 17120x (1973).

49 Zahir, S. A. C., Rembold, H., and Losert, E., German Patent 2,342,407 (1974); *Chem. Abstr.*, **81**, 113724e (1974).

50 Smith, A. C., Jr., Williams, J. L. R., and Unruh, C. C., U.S. Patent, 2,816,091 (1957); *Chem. Abstr.*, **52**, P 4369a (1958).

51 Munder, J., Ruckert, H., Steppan, H., Messwarb, G., and Lueders, W., (Kalle A-G), S. African Patent 6804,327 (1969); *Chem. Abstr.*, **71**, 86597a (1969).

52 Kalle, A-G., French Patent 1,573,500 (1969); *Chem. Abstr.*, **72**, 138327e (1970).

53 Kato, M., Ichijo, T., Ishii, K., and Hasegawa, M., *J. Polymer Sci., Part A-1*, **9**, 2109 (1971).

54 Atkinson, R. B., U.S. Patent 3,410,824 (1968); *Chem. Abstr.*, **70**, 20691z (1969).

55 Schellenberg, W. D., and Heinz, J., Farbenfabriken Bayer Akt.-Ges., German Patent 1,039,835 (1959); *Chem. Abstr.*, **55**, 7119f (1961).

56 Gunther, M. (Agfa Akt.-Ges.), German Patent 1,047,013 (1958); *Chem. Abstr.*, **55**, 8137f (1961).

57 Fox, C. J. (Eastman Kodak Company), British Patent 1,143,340 (1969); *Chem. Abstr.*, **70**, 92292t (1969).

58 Contois, L. E., and Merrill, S. H., French Patent 2,016,435 (1969); *Chem. Abstr.*, **74**, 93486d (1971).

59 Contois, L. E., and Specht, D. P., French Patent 2,022,355 (1970); *Chem. Abstr.*, **74**, 133060u (1971).

60 Yamaguchi, H., and Aoki, K., *Nippon Shashin Gakkai Kaishi*, **20**, 10 (1957); *Chem. Abstr.*, **53**, 107d (1959).

61 Panda, S. P., *Indian J. Technol.*, **9**, 387 (1971).

62 Panda, S. P., *Indian J. Technol.*, **11**, 356 (1973).

63 Koelbel, H., Manecke, G., and Guettler-Pimenidou, E., German Patent 2,256,961 (1974); *Chem. Abstr.*, **82**, 73919j (1975).

64 Ehrig, B., Mueller, E., and Mott, L., German Patent 2,013,414 (1971); *Chem. Abstr.*, **76**, 86380c (1972).

65 Delcarmen, M., Barrio, G., Barrio, J. R., Walker, G., Novelli, A., and Leonard, N. J., *J. Am. Chem. Soc.*, **95**, 4891 (1973).

66 Lense, F. T., Glover, C. A., and Markham, E. C., *Virginia J. Sci.*, **3**, 14 (1942); *Chem. Abstr.*, **36**, 3113^9 (1942).

67 Syamasundar, K., *Proc. Indian Acad. Sci.*, **59A**, 241 (1964).

68 Mitsubishi Petrochemical Co. Ltd., British Patent 1,128,090 (1968); *Chem. Abstr.*, **69**, 97342y (1968).

69 Hercules Powder Co., British Patent, 873,021 (1959); *Chem. Abstr.*, **56**, 6185h (1962).

70 Hayakawa, G., and Inoue, T., Japanese Patent 7107,386 (1971); *Chem. Abstr.*, **74**, 143332y (1971).

71 Phillip's N.V., Gloeilampenfabrieken, Netherland Patent 6,501,841 (1966); *Chem. Abstr.*, **66**, 16034n (1967).

72 Baeyens, P., and Krijl, G., *Trans. Inst. Metal. Finish*, **45**, 115 (1967); *Chem. Abstr.*, **67**, 104540a (1967).

Index